Genetic Editing for Healthcare

Advances in Disease Management

Genetic Editing for Healthcare

Advances in Disease Management

Bhupinder Singh, PhD
Rishabha Malviya, PhD
Sahil Lal, PhD
Sathvik Belagodu Sridhar, PhD
Editors

AAP APPLE ACADEMIC PRESS

First edition published 2026

Apple Academic Press Inc.
1265 Goldenrod Circle, NE,
Palm Bay, FL 32905 USA

760 Laurentian Drive, Unit 19,
Burlington, ON L7N 0A4, Canada

CRC Press
2385 NW Executive Center Drive,
Suite 320, Boca Raton FL 33431

4 Park Square, Milton Park,
Abingdon, Oxon, OX14 4RN, UK

Library and Archives Canada Cataloguing in Publication

CIP data on file with Canada Library and Archives

Library of Congress Cataloging-in-Publication Data

CIP data on file with US Library of Congress

ISBN: 978-1-77964-381-0 (hbk)
ISBN: 978-1-77964-382-7 (ebk)

DOI: 10.1201/9781779643827

About the Editors

Bhupinder Singh, PhD
Professor, Sharda School of Law, Sharda University, Greater Noida, India; Honorary Professor, University of South Wales, UK; Santo Tomas University Tunja, Colombia

Bhupinder Singh, PhD, is working as a Professor at the Sharda School of Law at Sharda University Greater Noida, India. He is also a Honorary Professor at the University of South Wales, United Kingdom (UK), and Santo Tomas University Tunja, Colombia. He has written three books, published 139 research papers, and presented 163 papers at international/national conferences and seminars, and has participated in more than 40 workshops, faculty development programs, and quality Improvement programs. He has taken over 25 courses at international universities, organized more than 59 events with international and national academicians and industry professionals, developed new courses, and served as editor-in-chief and co-editor for several journals. He has given talks at international universities and has been a resource person at international conferences around the world. His leadership, teaching, research, and industry experience spans more than 16 years. His research interests are health law, criminal law, research methodology, and emerging multidisciplinary areas as blockchain technology, the Internet of Things (IoT), machine learning (ML), artificial intelligence (AI), genome-editing, photovoltaic (PV) glass, Sustainable Development Goals, etc.

Rishabha Malviya, PhD
Associate Professor, School of Medical & Allied Sciences, Galgotias University, Greater Noida, Uttar Pradesh, India

Rishabha Malviya, PhD, is working as an Associate Professor in the Department of Pharmacy, School of Medical and Allied Sciences, Galgotias University, India. He has 12 years of research experience. His area of interest includes formulation optimization, nanoformulation, targeted drug delivery, localized drug delivery, and characterization of natural polymers as pharmaceutical excipients. He has authored more than 200 research and review papers for national and international journals of repute. He has to his credit 58 patents (19 grants, 38 published, one filed) as well as many publications in reputed national and international journals. He has received an Outstanding Reviewer Award from Elsevier. He has authored and edited nearly 50 books (Wiley, CRC Press/Taylor and Francis, Springer, River Publisher, IOP Publishing, and OMICS Publishing Group) and authored over 30 book chapters. His name has included in the World's Top 2% Scientists by Stanford University. He is a reviewer, editor, and editorial board member for more than 50 national and international journals of repute. He was an invited author for the magazine Atlas of Science as well as a pharma magazine dealing with B2B industry, Ingredient South Asia. He completed his BPharm from Uttar Pradesh Technical University and MPharm (Pharmaceutics) from Gautam Buddha Technical University, Lucknow Uttar Pradesh, India. His PhD (Pharmacy) work was in novel formulation development techniques.

Sahil Lal, PhD
Assistant Professor, School of Law, Galgotias University, India

Sahil Lal, PhD, is an Assistant Professor in the School of Law at Galgotias University, India. He has more than eight years of academic, industry, and research experience. He has been associated with Galgotias University since 2025. He is a distinguished legal academic and practitioner with an extensive background in both academia and the legal industry. He has also served as guest faculty at the Faculty of Law at the University of Delhi in 2022. Dr. Lal holds a PhD from Sharda University, with his thesis titled "Intellectual Property Rights in The Era of Artificial Intelligence: An Analytical Study." He earned his LLM from Gujarat National Law University, Gandhinagar, and graduated with a B.A. LL.B. (Hons) from Dr RML National Law University, Lucknow, following his schooling at The Scindia School, Gwalior. He has qualified UGC-NET & JRF.

Sathvik Belagodu Sridhar, PhD
Associate Dean and Professor, Department of Clinical Pharmacy and Pharmacology, RAK Medical and Health Sciences University, Ras Al Khaimah, United Arab Emirates

Sathvik Belagodu Sridhar, PhD, is currently working as an Associate Dean and Professor at the RAK College of Pharmacy (RAKCOP), RAK Medical and Health Sciences University, located in Ras Al Khaimah, United Arab Emirates. He received an FIP (International Pharmaceutical Federation) fellowship for higher training in renal clinical pharmacy at the Royal Adelaide Hospital and Queen Elizabeth Hospital in Adelaide, South Australia. Dr. Sridhar has been honored with a prestigious teacher award by the Society of Pharmaceutical Education & Research (SPER). He has also received training in pharmacovigilance at the Uppsala Monitoring Centre in Uppsala, Sweden. He has 25 years of experience in teaching and practicing clinical pharmacy in India, Malaysia, and the UAE. His research interests include medication safety, pharmacovigilance, pharmacoepidemiology, and renal and psychiatry clinical pharmacy. Dr. Sridhar has published more than 75 research papers in reputable national and international journals, along with book chapters. He has presented more than 75 research papers at national and international conferences and has been honored with many awards. He serves as a reviewer for various national and internal pharmacy practice-related journals. Dr. Sridhar completed his PhD in Clinical Pharmacy from Rajiv Gandhi University of Health Sciences, Bangalore, India.

About the Book

The book *Genetic Editing for Healthcare: Advances in Disease Management* delves into the revolutionary role of genetic editing in transforming the landscape of healthcare. With the advent of precision medicine, genetic editing techniques such as Clustered Regularly Interspaced Short Palindromic Repeats–CRISPR-associated protein 9 (CRISPR-Cas9) and other genome-editing tools have become powerful methods for treating, preventing, and potentially curing various genetic diseases and disorders. This book provides a comprehensive exploration of how genetic editing technologies are applied in healthcare, their current developments, ethical concerns, and future potential in advancing disease management. In recent years, advancements in genomics have provided unprecedented insights into the human genome, unlocking new possibilities for medical interventions that were once considered science fiction. The ability to directly alter the DNA of living organisms represents a leap in scientific innovation that is pushing the boundaries of traditional medical treatment. The book thoroughly examines the impact of these cutting-edge genetic editing techniques on specific healthcare areas, ranging from hereditary diseases like cystic fibrosis and sickle cell anemia to chronic conditions such as cancer and heart disease.

This work also discusses the precision and personalization that genetic editing brings to medical treatment. By targeting specific mutations, genetic editing tools allow healthcare professionals to treat diseases at their root causes, offering hope for more effective and long-lasting treatments. As researchers continue to refine these technologies, their applications are expanding—from gene therapies to advancements in organ transplantation, drug discovery, and regenerative medicine.

The book goes beyond the scientific and technical details, addressing the ethical, legal, and social implications of genetic editing. These concerns include questions about the long-term effects of gene editing, potential misuse in non-therapeutic applications, and the equitable distribution of these technologies across global populations. By tackling these broader issues, the book aims to foster a balanced understanding of both the promise and the challenges associated with genetic editing in healthcare.

The role of genetic editing in the future of healthcare is profound, but it also presents challenges that must be navigated with care and foresight. Ultimately, the book argues that while genetic editing holds extraordinary potential for improving human health, a thoughtful and ethical approach is essential to ensure its benefits are realized responsibly and equitably.

Contents

Contributors

Saquib Ahmed
Sharda School of Law, Sharda University, Greater Noida, Uttar Pradesh, India

Manmeet Kaur Arora
Sharda University, Greater Noida, Uttar Pradesh, India

Kozma Dorottya Edina
University of Pannonia, Veszprém, Hungary

Ashima Jain
School of Law, Manipal University Jaipur, Jaipur, Rajasthan, India

Christian Kaunert
Dublin City University, Dublin, Ireland; and University of South Wales, Pontypridd, Wales, United Kingdom

Sahil Lal
Assistant Professor, School of Law, Galgotias University, India

Rishabha Malviya
Galgotias University, Greater Noida, Uttar Pradesh, India

Fatima Zahra Ouariach
Abdelmalek Essaadi University, Tétouan, Morocco

Soufiane Ouariach
Abdelmalek Essaadi University, Tétouan, Morocco

Shayut Pavapanunkul
Universitas Muhammadiyah Sidenreng Rappang, Sidenreng Rappang, South Sulawesi, Indonesia

Anjali Raghav
Sharda School of Law, Sharda University, Greater Noida, Uttar Pradesh, India

Richa Raghav
Sharda University, Greater Noida, Uttar Pradesh, India

Bhupinder Singh
Sharda School of Law, Sharda University, Greater Noida, Uttar Pradesh, India; University of South Wales, UK; Santo Tomas University Tunja, Colombia

Thanaporn Sriyakul
Association of Legal and Political Studies, Bangkok, Thailand

Abbreviations

6-MP	6-mercaptopurine
ADME	absorption, distribution, metabolism, and excretion
ADR	adverse drug reactions
AI	artificial intelligence
ALL	acute lymphoblastic leukemia
APHIS	animal and plant health inspection service
AV	adenovirus
AZA	azathioprine
CAR	chimeric antigen receptor
Cas	CRISPR-associated
CBD	convention on biological diversity
cDNA	complementary DNA
CF	cystic fibrosis
CFIA	Canadian Food Inspection Agency
COVID-19	coronavirus disease 2019
CRISPR	clustered regularly interspaced short palindromic repeats
CRS	cytokine release syndrome
ctDNA	circulating tumor DNA
CYP	cytochrome P450
DNA	deoxyribonucleic acid
DSB	double-strand break
DTC	direct-to-consumer
EBI	European Bioinformatics Institute
EGFR	epidermal growth factor receptor
EGFR-TKI	epidermal growth factor receptor-tyrosine kinase inhibitor
EHR	electronic health record
EMA	European Medicines Agency
EPA	Environmental Protection Agency
EU	European Union
FDA	Food and Drug Administration
FIFRA	Federal Insecticide, Fungicide, and Rodenticide Act
GA4GH	Global Alliance for Genomics and Health
GDPR	general data protection regulation
GINA	Genetic Information Nondiscrimination Act
GIST	gastrointestinal stromal tumor
GM	genetically modified

GMO	genetically modified organism
GO	gene ontology
gRNA	guide RNA
GWAS	genome-wide association studies
H3Africa	human heredity and health in Africa
HDR	homology-directed repair
HFEA	Human Fertilization and Embryology Act
HGP	human genome project
HIV	human immunodeficiency virus
HPC	high-performance computing
HSC	hematopoietic stem cell
HTS	high-throughput screening
IBD	inflammatory bowel disease
IPR	intellectual property right
MDR	multidrug-resistant
ML	machine learning
NCD	non-communicable disease
NGS	next-generation sequencing
NHEJ	nonhomologous end joining
NIH	National Institutes of Health
NSCLC	non-small cell lung cancer
PCI	percutaneous coronary intervention
PCR	polymerase chain reaction
PCT	patent cooperation treaty
PKU	phenylketonuria
SCNT	somatic cell nuclear transfer
SECURE	sustainable, ecological, consistent, uniform regulatory evaluation
sgRNA	single guide RNA
SMA	spinal muscular atrophy
SNP	single nucleotide polymorphism
SSRI	selective serotonin reuptake inhibitor
TALEN	transcription activator-like effector nucleases
TCGA	The Cancer Genome Atlas
TKI	tyrosine kinase inhibitor
TMB	tumor mutational burden
TN	targeted nutrition
TPMT	thiopurine S-methyltransferase
tracrRNA	trans-activating crRNA
TRIPS	trade-related aspects of intellectual property rights
UGT	UDP-glucuronosyltransferase
USDA	U.S. Department of Agriculture

WGS	whole-genome sequencing
WHO	World Health Organization
WTO	World Trade Organization
ZFN	zinc finger nuclease

Foreword

The book *Genetic Editing for Healthcare: Advances in Disease Management,* edited by Dr. Rishabha Malviya and his team, is a thorough, multidisciplinary examination of how genetic editing is revolutionizing medicine and providing fresh hope in the battle against complex illnesses. The book delves into the governance and regulatory frameworks required for responsible gene editing—namely, Clustered Regularly Interspaced Short Palindromic Repeats (CRISPR) technology—emphasizing the importance of legislation in protecting human rights and fostering responsible innovation. It also explores the digital infrastructure required for the successful adoption of gene-editing technologies, as well as solutions for bridging the digital gap and improving healthcare access.

The book progresses into the specific applications and limitations of CRISPR in genetic editing, addressing topics such as plant tolerance to environmental pressures and complex approaches to human genome modification. The insights into the biosafety challenges of plant genome editing offer an in-depth understanding of the interconnectivity between human health and environmental sustainability, emphasizing the significance of careful risk assessment.

Dr. Malviya and his team have compiled an extensive range of chapters on the ethical, legal, and societal consequences of gene editing, with a focus on human rights concerns and the legal frameworks that must adapt to incorporate the impact of gene editing. The book also highlights advances in genomics for drug discovery, which use bioinformatics and artificial intelligence (AI) to create more personalized and precise medications. *Genetic Editing for Healthcare: Advances in Disease Management* is an essential resource for healthcare experts, researchers, clinicians, and biomedical scientists involved in the future of healthcare and genetic technology. As you engage with these chapters, you will find yourself at the intersection of science, ethics, and innovation—an intersection where the future of healthcare is being shaped.

Happy Reading and Best Wishes!

—Dhruv Galgotia, PhD
CEO, Galgotias University,
Greater Noida, Uttar Pradesh, India

Preface

The scientific community and the general public have been intrigued by the rapid evolution of the discipline known as genetic editing. What was once science fiction has now become scientific fact. With methods of gene editing like Clustered Regularly Interspaced Short Palindromic Repeats (CRISPR), the future of healthcare has arrived. As a result, there has been enormous enthusiasm, giving rise to ongoing curiosity and debate—and this is the backdrop from which this book originated. Healthcare has traditionally focused on treating the symptoms of diseases, but what genetic editing allows us to do is target disease at its source: the genome.

Ultimately, the purpose of this book is not just to capture what has been learned about genetic editing, but also to make that knowledge accessible to everyone. This book is intended for a wide audience, including students, healthcare professionals, policymakers, and informed laypeople interested in understanding how genetic editing could influence human health. It also aims to draw attention to the increasing importance of ethical, legal, and societal discussions, without which this science should not advance. The initial chapters lay the groundwork for understanding the molecular foundations and major scientific frameworks behind these technologies, with special focus on CRISPR-associated protein 9 (CRISPR-Cas9), transcription activator-like effector nucleases (TALENs), and zinc finger nucleases (ZFNs), which are now among the principal tools. The book also enlightens readers with detailed chapters on their mechanisms of action at the molecular level, and how effectively they can assist biological machinery in overcoming genetic failures.

After that, it discusses the practical uses of genome editing in our healthcare repositories. Through stories of those newly armed with gene-modifying tools to tackle life-threatening genetic diseases—including case studies on its deployment in cancer treatment settings—the book illustrates how patients are now benefitting from the realization of gene editing in disease therapy. The promise is immense, but so are the challenges and complexities of gene-editing therapies. This contribution emphasizes the need for consensus on the appropriate regulatory frameworks and ethical standards that must be met before genetic editing is widely implemented. This topic is addressed in two parts.

We are at a pivotal moment in deciding how society will ethically and responsibly integrate genetic editing into everyday healthcare. These developments are mind-boggling, but they also carry very serious implications. The final fifth of the text explores what lies ahead for genetic editing in healthcare and offers insights into emerging tools that may soon become available—potentially transforming this field even further. Whether it is advances in regenerative medicine, organ transplantation, or gene therapies, current developments are only scratching the surface of what is achievable. This book aims to stimulate interest, broaden the reader's understanding of these issues, and contribute to a meaningful dialogue about the principles guiding genetic editing in healthcare. If done correctly, genetic editing holds the potential to become one of humankind's most transformative advances in medicine. The journey is far from smooth, with many obstacles along the path, but we must move forward—taking all necessary safety measures and remaining ever cautious.

CHAPTER 1

Gene Editing Governance and Genetic Engineering: A Framework for Regulating CRISPR Technology

CHRISTIAN KAUNERT[1,*] and RICHA RAGHAV[2]

[1]*Dublin City University, Dublin, Ireland; and University of South Wales, Pontypridd, Wales, United Kingdom*

[2]*Sharda University, Greater Noida, Uttar Pradesh, India*

*Corresponding author

ABSTRACT

The advancement of CRISPR-Cas9 technology over the past few years has revolutionized gene editing, providing extensive tools for facile genetic engineering, and these developments also pose considerable ethical, legal, and social issues that demand a broader governance framework. This framework incorporates a broad range of stakeholder perspectives to help guide the ethical challenges that gene editing presents, while simultaneously enabling necessary innovation in medical application through provision of enforceable ethical norms. CRISPR technology appears in the gene editing process and has revolutionized it, providing great potential benefits in medicine, but also agriculture and biotechnology. While advancing rapidly, this still nascent technology is presenting complex ethical, safety, and regulatory questions. Also, investigate the state of gene editing governance and argue that such oversight mechanisms should be implemented based on characteristics shared across typical human genome editing cases, especially germline applications.

Genetic Editing for Healthcare: Advances in Disease Management.
Bhupinder Singh, Rishabha Malviya, Sahil Lal, and Sathvik Belagodu Sridhar (Eds.)

DOI: 10.1201/9781779643827-1

1.1 INTRODUCTION

CRISPR-Cas9 technology was developed and quickly became a vital tool in the field of genetic engineering for editing genes with incomparable precision and efficiency. This tool can, if treated with the respect it deserves, lead to breakthrough developments in many fields like medicine, agriculture, and biotechnology [1]. Also, more far-reaching innovations in the realm of gene editing are on the horizon and call for a comprehensive governance framework due to critical ethical, legal, and societal issues raised by them. But as scientists and policymakers alike mull the disruption cryptocurrencies could bring, there is growing demand for those regulations [2]. CRISPR technology, which lets scientists make precise changes to an organism's DNA through gene editing, could bring cures for genetic-based disorders along with increased agricultural yield. But these advancements bring massive ethical challenges. The new technology also raises the possibility of germline editing, changing genes in human embryos that will be passed on through generations, and that has stirred debate over "designer babies" and genetic enhancement [3]. These worries highlight the requirement for not only a governance framework that ensures safety and efficacy, but also that concerns ethics principles and societal values. At present, the governance systems that regulate activities in gene editing are often piecemeal and disjointed between countries or regions. While the regulations on germline editing have been strict in some countries, others allow it when certain conditions are met. This lack of consistency between laws can result in "scientific tourism," whereby individuals seek treatments in jurisdictions with low levels of restrictions, possibly exposing them to untested or unsafe procedures [4].

They fail to keep up with the tremendous pace at which technologies evolve, leaving gaps that can be exploited. Against this backdrop, international entities such as the World Health Organization (WHO) are initiating conversations around forming overarching governance structures for human genome editing [5]. The WHO's Expert Advisory Committee on Developing Global Standards for Governance and Oversight of Human Genome Editing has underscored the need to have guidelines that address scientific, ethical, social, and legal dimensions [6]. They hope to nail down a list of recommendations that can be flexible enough for different areas, and also ensure public trust and accountability. Accessible participation in the decision-making of the multi-stakeholders is one of the most important characteristics of a strong governance. It discusses how scientists, ethicists, policymakers, and members of the general public can work together to establish these acceptable boundaries for gene editing applications [7]. This stakeholder-led approach not only serves as a demonetization tool during the decision-making process but also nurtures collective stewardship in navigating the ethical environment about gene editing technologies.

These are assessed within the wider context of international initiatives such as that of the WHO, in conjunction with national regulatory approaches; it is a picture which appears fragmented and often lacking in clarity and consistency. This research chapter reviews the governance of gene editing at present, fills in regulatory gaps, and proposes a complete system to control the development of CRISPR-Cas9 technology. The goal of this framework is to maintain a balance between innovation and ethics, which will enhance public trust and minimize potential risk [8].

1.2 INTERNATIONAL PERSPECTIVES ON GENE EDITING GOVERNANCE

The rapid development of gene editing technologies, most notably the CRISPR-Cas9 technology, has raised questions about the need for frameworks to govern their application and use at a global scale [9]. Human genome editing has brought an array of challenges for national and international policymakers, rife with ethical, legal, and social implications [9]. The explosive growth in CRISPR technology of late could reshape entire industries, from medicine to agriculture to environmental remediation. On the other hand, these exciting new possibilities also present major ethical, safety, and regulatory hurdles; practical challenges that ultimately hinge on responsible balancing. Balancing the innovation potential of CRISPR with necessary safeguards that seek to protect public health, safety, and moral and ethical standards is not easy and requires a holistic governance framework. The multi-tiered regulatory structure proposed is the most important step to unravel the tangle of gene editing governance and, importantly, avoids dealing with all gene editing technologies as equals in regulatory terms. The framework blends local, national, and international sticks with carrots specifically suited to each community so that the need for regulations catering to the lowest common denominator is reduced, while global cooperation is maximized. Managing these gene editing technologies with local regulations will allow for more localized concerns to be addressed and allows stakeholders a say in how these tools will be applied. International agreements allow for cooperation and experience delivering best practices when it comes to border necessity but have frameworks that are developed on the national level. Its necessity is amplified by the fact that many of the topics impacting synthetic biology and biotechnology as a whole are global issues.

The speed with which science in advanced gene editing has progressed is unmatched, and regulation will need to be capable of adapting quickly to new findings and changing ethical situations. A system of regular checks and updates should be in place so the spirit and effectiveness of existing policies are maintained. This same adaptability is crucial in being proactive about governance, identifying, and solving things before they become social challenges. Public engagement and education are also important elements of the model being proposed. A diverse array of stakeholders, including scientists, ethicists, the public, and policymakers, are necessary to ensure that CRISPR technology is completely understood. Regulatory bodies can further enable informed discussions on the possible benefits and risks of gene editing by establishing avenues for discussion forums and educational efforts. This not only increases transparency but also builds public confidence in the regulatory process. A well-informed public is necessary to have a lively discussion on the gene editing front that reflects various types of social activities. The more important point in this framework is the creation of ethical review boards independent from their host organization. Our approach depends on the advice of multidisciplinary experts who sit on such boards, which may allow in-depth deliberation of complex ethical dilemmas that are presented by CRISPR applications. The purpose of these committees would be to vet or review proposed research against adherence to fairly established ethical benchmarks by examining any societal implications as such. This integration of diverse knowledge into the evaluation process would ensure higher quality oversight and more ethical behavior in research.

1.3 NEED FOR GOVERNANCE

The use of CRISPR has made tremendous ethical noise, as on germline editing and thus transmissible genetic modifications. Chinese scientist He Jiankui sparked global controversy when he claimed to have genetically modified twin girls in 2018, highlighting the urgent need for an international regulatory system [10]. His work, showing human genome editing was possible and could produce apparently healthy babies, was condemned by many and raised alarm about the dangers of uncontrolled gene experimentation on humans [11]. The news of such incidents has prompted discussion about gene editing governance nationally and internationally [12].

1.3.1 INTERNATIONAL INITIATIVES

The WHO is a major player on governance frameworks, having established an Expert Advisory Committee on Developing Global Standards for Governance and Oversight of Human Genome Editing. The committee will identify the scientific, ethical, social, and legal challenges posed by human genome editing [13]. The WHO framework values stronger regulatory bodies that can be replicated across settings, building confidence in the public and for transparency. Developing governance models overseas is also underway by a range of different international health and science organizations [14]. These efforts include discussions at scientific conferences and summits that bring together key regulatory agencies from around the world to harmonize standards [15]. The aim is to generate a consensus towards enhancing the safety and ethical aspects for innovation in gene editing technologies. The legislative landscape around gene editing is strikingly patchy, with different nations conceiving different societal values and risk/development priorities for the technology [16]. The hodgepodge of national regulatory approaches can also create a fractured landscape that undermines collaboration and invites "scientific tourism," where patients seek treatments in jurisdictions with less strict regulations [17].

1.3.1.1 UNITED STATES

The regulations of gene editing in the United States reflect a federalist system whereby power is dispersed with respect to three agencies: the Food and Drug Administration (FDA), the U.S. Department of Agriculture (USDA), and the Environmental Protection Agency (EPA) [18]. Every agency has its own conformance and restrictions about the gene editing communities that includes [20]:

1. **FDA:** The FDA regulates gene editing through the lens of drug approvals. Human gene therapies are defined as biologic products and, thus, must undergo extensive tests to ensure safety and efficacy before testing in clinical trials. But there is no comprehensive

federal law that applies directly to germline editing [21]. In reality, the amendment has been in place since 2016 that bans the FDA from reviewing any applications for a clinical trial that proposes human germline genetic modification. This restriction may be appropriately conservative, given the nascent state of germline editing, but it simultaneously poses difficulties for investigators who wish to experiment with potential treatments [22].

2. **USDA:** The USDA regulates farm applications of gene editing through its Animal and Plant Health Inspection Service (APHIS). In 2018, APHIS finalized the SECURE (sustainable, ecological, consistent, uniform regulatory evaluation) rule, which exempts certain gene-edited plants from regulatory review if they could have been developed through conventional breeding [23]. This is a product-focused approach where the focus is on product characteristics instead of how it was made, and hence results in a more simplified regulatory pathway for low-risk modifications [24].
3. **EPA:** For gene-edited crops that are insecticides, produced through other-gene pesticidal regulation applied under the Federal Insecticide, Fungicide, and Rodenticide Act (FIFRA). Prior to issuing approval, the agency evaluates these products for potential risk to human health and the environment [25]. This fragmented American regulatory climate adds burden to the shoulders of researchers and developers, who face different standards and requirements from the plethora of respective agencies. Moreover, the absence of federal guidelines for germline editing means that researchers who are interested in this realm cannot move forward with much confidence [26].

1.3.1.2 EUROPEAN UNION (EU)

The European Union, in contrast, has been more hesitant about gene editing than the United States, as the regime of the EU is known by its official standards in safety and ethics for:

1. **Regulatory Framework:** The European Medicines Agency (EMA) subjects gene therapies to an extensive safety evaluation before its clinical application is approved [27]. The EU has long

been wary of some forms of GM (especially germline editing), on which it takes a precautionary approach [28].

2. **Germline Editing Prohibition:** Numerous EU member states have enacted laws that specifically forbid any intervention in the human germline [30]. This ban is rooted in moral agonizing's about what introducing gene editing to humans might mean for the future and broader implications of a bravado scramble to patch up our physiognomy using genetic seam stressing [31].

3. **Public Sentiments:** In Europe, public sentiment on genetic engineering technology development is much more skeptical than in other regions of the world; as a result, For example, politicians have to act stricter in terms of legislation against GM food. This is a cautious, measured way to ensure public health and standards in research ethics are maintained [32].

1.3.1.3 CHINA

China cracked down on unauthorized gene editing following international scrutiny post the He Jiankui scandal such as:

1. **Strict Regulation:** In response to He Jiankui's claim of having edited the germline genome of embryos, China has begun working on tighter regulations concerning human gene editing [33]. The Chinese government highlighted its regulatory system in line with international standards and ethical requirements.

2. **Focus on Ethics:** New regulatory guidelines in China center on ethical concerns, mandating that researchers must obtain approvals from ethics committees to conduct human genome editing research [34]. This transition is to bring back the faith of general people in scientific research on the one hand and to conform with ethics norms on the other.

3. **International Collaboration:** As highlighted in the COS special section, China clearly acknowledges that it needs to harmonize its regulatory system with international standards if it is going to be able to participate in joint scientific undertaking [35]. Through the introduction of tighter governance protocols, China wants to present itself as a responsible global biotech player.

1.4 IMPLICATIONS OF FRAGMENTED REGULATORY APPROACHES

These different national approaches to gene editing governance make for a patchier international tapestry, which is not without its issues such as:

1. **Scientific Tourism:** This could lead people who are in the market for experimental treatments to travel to countries with looser regulations and thereby subject themselves to unproven or unsafe interventions [36]. The point that arises here is the ethical one regarding informed consent and patient safety.
2. **Research Collaboration:** Regulations that become fragmented can prevent international collaboration between researchers and institutions if contrarian standards impair any joint projects or data-sharing [37].
3. **Public Trust:** A patchwork of regulations could also undermine public trust in gene editing technologies [38]. Transparent and comprehensive governance systems are crucial in promoting trust and trust building among all stakeholders, including patients, researchers, and policy makers [40].

1.5 ETHICAL CHALLENGES IN GENE EDITING TECHNOLOGIES

1.5.1 DESIGNER BABIES AND GENETIC ENHANCEMENT

Ethical issues surrounding gene editing include the possibility of creating "designer babies." Gene editing involves editing human embryos using genetic modification techniques to choose or confer particular qualities in humans, like looks, IQ, and health. The ethical considerations here are strong as [41]:

1. **Equity and Access:** Genetic editing raises concerns regarding social inequality due to the possibility of manipulating genetic traits [42]. If these upgrades are only accessible to those who are already wealthy, it could cascade us into a caste system of genetic haves and have-nots that will only further social inequalities.
2. **Approval and Autonomy:** Germline editing not only causes the person to be altered, but future generations as well, who cannot agree to the changes that are done in their genomes. It does,

however, raise some questions on where parental authority ends and the rights of the children to inherit their unmodified genome begin [43].

3. **Unintended Consequences:** The long-term effects of genetic modifications are a concern. If certain traits were modified, there is a concern that health or behavior risks could occur in other unanticipated ways, and who would be responsible under those circumstances: parents or scientists [44]?

1.5.2 ECOLOGICAL RISKS

The release of gene-edited organisms into ecosystems poses major ecological risks that are appropriate to address:

1. **Diversity:** Gene editing can unintentionally change the metabolic pathways of an organism, with potential consequences for existing food webs and ecosystems [45]. This could, for instance, result in changing pollinator-sensitive fatty acid profiles of crops and consequently reverberate through the entire pollination cascade.
2. **Gene Flow and Hybridization:** There is a risk that gene-edited plants could hybridize with wild relatives, polluting natural populations with transgenically altered genes [46]. This gene flow has the potential to disrupt ecosystems and biodiversity by out-competing native species or swamping their genomes with alien genes [47].
3. **Environmental Persistence:** Gene-edited organisms could become more persistent in the environment than previously estimated, which could lead to longer-term ecological effects [48]. Their unrestrained proliferation potential warrants comprehensive risk assessments as a precursor to field release [49].

1.5.3 BIOSECURITY THREATS

Biosecurity concerns regarding gene editing technologies stem from the potential for misuse such as:

1. **Dual Use Dilemma:** Given that gene editing tools are dual use, that is, the technology can be used for benevolence (medical

therapies) as well as malevolence [50]. CRISPR tech is easier to get these days, and that means a better chance that people who want to use CRISPR for things they're really not supposed to be using it for.

2. **Pathogen Modification:** Gene editing could be used to build more virulent strains of pathogens or make them resistant to treatments. If those modifications were to get out of the lab or be deliberately released into the environment, this is a real risk to public health and public safety [51].
3. **Regulatory Gaps:** Formal laws and regulations may lag behind new technologies, leaving gaps in which uncertainty about the application of legal principles can be exploited [52]. Without effective global oversight of dual-use research, it has been difficult to provide guidance on how beneficial applications can be promoted whilst misuse is prevented.

1.6 RESEARCHERS AND POLICYMAKERS: OVERVIEW ON GUIDANCE

Many of these (ethical) frameworks are likely to fail in providing the necessary form of guidance for researchers and policymakers who make their way through gene editing labyrinths:

1. **Changing Technological Landscape**: Novel gene editing technologies advance so fast that existing ethical standards might not be capable of dealing with new problems as they arise [53]. Regulations need to accommodate new technologies, and there must be an ongoing dialogue between scientists, ethicists, and policymakers [54].
2. **Comprehensive Risk Assessments**: There is an increasingly recognized need for comprehensive risk assessments that address the full spectrum of acute and chronic effects on human health and the environment, including unintended consequences [55]. Regulatory frameworks should include these assessments to guarantee the safe use of gene editing technology.
3. **Public Engagement:** Involving a wide variety of stàkeholders in discussions about gene editing, from scientists and ethicists

to policymakers and the public, will certainly play an essential role in shaping the ethical framework. Public input can help guide the discussion toward policies that represent societal values and communities regarding genomic tools [56].

1.7 PROPOSED REGULATORY FRAMEWORK

The new technologies for gene editing, especially CRISPR, are developing at a rapid pace and call for a thorough regulatory system to control them in order to use them safely and ethically. The framework needs to incorporate all the aspects of the tiered multi-level and adaptive system in a way that makes it inclusive, public facing, and addresses ethical issues [57].

1.7.1 MULTI-TIERED REGULATORY STRUCTURE

Cross-border transactions involving multitudes of local regulators, a multi-tiered regulatory structure would be needed due to differing levels of financial sophistication and experience observed in different emerging markets [58]. Addressing the multiplicity of contexts in which gene editing technologies are applied, however, will require a multi-tiered regulatory framework. Its structure involves a cocktail of local, national, and global ingredients such as:

1. **Local Regulations:** They can be used to address particular concerns in the community and consider cultural values. In other words, it means that agricultural or healthcare processes are clear to the community and are adding value, at a technical level, to speed up the process of vaccination [59]. Community accountability can allow for more individualized responses, based on the unique needs and ethical considerations of specific groups.
2. **National Frameworks:** Rules passed at the national level establish a common framework on which local laws can build. Such frameworks ought to establish a floor of safety, effectiveness, and ethical adherence that practices of gene editing do not fall below. They may also provide a foundation for research funding and public health initiatives [60].
3. **International Agreements:** Since issues related to gene editing are trans-boundary, international cooperation is a major requirement.

By entering into global agreements, countries can help align their rules and oversight structures at a high level that captures common ethical principles to support international research collaborations. But these kinds of deals can be useful in preventing people from resorting to "scientific tourism," where patients travel overseas for therapies with lower regulatory hurdles [61].

1.7.2 REGULATORY ADAPTATION MECHANISMS

Incorporate adaptive regulatory mechanisms that allow flexibility in response to new scientific findings and societal concerns into the proposed framework:

1. **Routine Evaluations**: Oversight of the relevance and effectiveness of existing regulations must take place on a routine basis. It includes keeping an eye on developments in gene editing technologies and the extent of their risk for public health and safety [62].
2. **Timely Update:** Requiring the ability to update the regulations as new ethical issues and new tech become relevant. Regulations, for instance, may need to change in response to new CRISPR variants if they show increased specificity and decreased off-target effects [63].
3. **Stakeholder Feedback:** Regularly reviewing regulatory frameworks with the input of stakeholders (researchers, ethicists, and members of the public) can help to make them more adaptable. Feedback systems can assist in the detection of weak or unclear areas in need of reform [64].

1.7.3 PUBLIC ENGAGEMENT AND EDUCATION

Public engagement is an important part of informed decision-making and establishing trust in how gene editing will be governed as [65]:

1. **Establishing Forums:** Forums to discuss the myriad of perspectives on the implications surrounding gene editing technologies. For example, they may provide an opportunity for dialogue between scientists, policymakers, ethicists, and community members [66].
2. **Educational Initiatives**: Public education campaigns to educate citizens about gene editing and the risks-to-benefits relationship [66]. Educating individuals on the basics of

scientific process and the underlying bioethical principles can help these initiatives to elicit informed participation in societal debate about governing a transformational technology like gene editing [58].

3. **Public Consultations:** Requiring regular public consultations to provide opportunities for community input on proposed regulations or gene editing research projects [66]. By involving the public in these discussions, researchers can help create transparency and trust between them and the community [67].

1.7.4 ETHICAL REVIEW BOARDS

Formation of separate ethical review boards for appropriate advice for intricate issues in gene editing:

1. **Diverse Expertise:** These boards should also be composed of a broad range of experts, including ethicists, scientists, legal scholars, and community representatives, to evaluate proposed research projects/interventions using CRISPR technology holistically.
2. **Evaluation Criteria:** The boards would apply agreed-upon ethical standards, within which they will also consider anticipated societal implications to assess research proposals. They will also have the key responsibility of making sure research is based on community values and ethics.
3. **Advisory Role:** In addition to the oversight function described above, these boards could serve as a mechanism for advising policymakers on emerging and related issues of gene-editing technologies [66].

1.8 CONCLUSION

The success of a regulatory framework will come down to its flexibility in the face of rapidly changing gene editing technologies and its ability to respond adequately to public concerns. It is also very important that new regulations are written in a way which does not hinder innovation, but most importantly, promotes the responsible research and application of CRISPR technologies. Creating an environment in which scientific progress can be sustained with strong ethical oversight would allow society to seize the potential benefits of gene-editing technologies that promise to improve health, enhance food security, and contribute solutions to environmental challenges.

1.9 FUTURE DIRECTIONS

With the inevitable use of CRISPR technology, regulating it vertically from a governance perspective is not only necessary but also inescapable. With regard to the incoming biotechnological era in which we are now on the border, responsibility, and ethics ought to be at our minimum threshold for exploration of what CRISPR can do. It should be a guide for future governance in the light of rapidly shifting scientific terrain, guaranteeing gene editing's potential transformability that could benefit society overall. Management of gene editing technology (e.g., CRISPR) must strike a balance between innovation and preserving public health, ensuring environmental equilibrium, and guaranteeing that ethical considerations are upheld. The biomarker is said to belong to a suitable class of genetic properties, according to the model-based approach for detecting spurious correlations. As we enter this brave new world of biotech, it is essential to establish strong governance mechanisms that align with our shared values and desires for a responsible genetic engineering future. Responsible governance will not only be essential to realizing the benefits of gene editing; as we chart our course through an ever more intricate map of scientific knowledge, it is one way of engaging the deep ethical contradictions it raises.

KEYWORDS

- **CRISPR-Cas9**
- **embryos**
- **gene editing**
- **genes**
- **genetic engineering technology**
- **human germline**
- **stakeholder**

REFERENCES

1. Stasi, A., & Thongpravati, O. (2024). Biotechnology innovation in do-it-yourself (DIY) gene editing: A call for a new regulatory framework. *OBM Genetics, 8*(2), 1–1.
2. Svingen, M., & Jahren, L. (2024). Making space for CRISPR: Scientists' translation work to make gene editing a legitimate technology. *Science and Public Policy, 51*(1), 15–27.
3. Clark, L. F., & Hobbs, J. E. (2024). Governing the unknown: Regulating future technologies. In *International Regulation of Gene Editing Technologies in Crops: Current Status and Future Trends* (pp. 79–94). Cham: Springer Nature Switzerland.
4. Clark, L. F., & Hobbs, J. E. (2024). Introduction: Why study the governance of gene edited agrifoods? In *International Regulation of Gene Editing Technologies in Crops: Current Status and Future Trends* (pp. 3–14). Cham: Springer Nature Switzerland.
5. Pandos, O. C. (2024). Traversing uncharted territory? The legislative and regulatory landscape of heritable human genome editing in Australia. *Federal Law Review, 52*(1), 75–102.
6. Munawar, N., Ahsan, K., & Ahmad, A. (2024). CRISPR-edited plants' social, ethical, policy, and governance issues. In *Global Regulatory Outlook for CRISPRized Plants* (pp. 367–396). Academic Press.
7. Singh, B., & Kaunert, C. (2024). Adventure in high altitude of mountainous topographies and health impacts: Lensing tourism sustainability via reducing ecological and sociocultural footprint and health emergency and medical assistance management. In G. Meraj, S. Hashimoto, & P. Kumar (Eds.), *Navigating Natural Hazards in Mountainous Topographies. Disaster Risk Reduction* (p. 15). Cham: Springer. https://doi.org/10.1007/978-3-031-65862-4_15 (accessed on 5th January 2026).
8. Singh, B., Singh, A., Kaunert, C., Arora, M. K., Lal, S., & Ravesangar, K. (2025). Unleashing the ethical and legal implications of e-business revolution: Consumer privacy and security concerns in phase of digital disruption. In H. Taherdoost, G. Drazenovic, M. Madanchian, I. Khan, & O. Arshi (Eds.), *Business Transformation in the Era of Digital Disruption* (pp. 157–180). IGI Global. https://doi.org/10.4018/979-8-3693-7056-8.ch006 (accessed on 5th January 2026).
9. Singh, B., & Kaunert, C. (2025). Innovative business models in digital and virtual reality for the gig economy: Space-age sustainable development projecting smart business practices. In H. Taherdoost, G. Drazenovic, M. Madanchian, I. Khan, & O. Arshi (Eds.), *Business Transformation in the Era of Digital Disruption* (pp. 127–156). IGI Global. https://doi.org/10.4018/979-8-3693-7056-8.ch005 (accessed on 5th January 2026).
10. Singh, B., Malviya, R., & Kaunert, C. (2024). Elevating workplace sustainability for employees lensing mental health advancements: Runway for future ready healthcare services projecting SDG 3 (good health and well-being). In E. Shaikh (Ed.), *Impact of Corporate Social Responsibility on Employee Wellbeing* (pp. 285–310). IGI Global. https://doi.org/10.4018/979-8-3693-3470-6.ch014 (accessed on 5th January 2026).
11. Singh, B., Kaunert, C., & Malviya, R. (2024). Featuring machine learning in municipal solid waste management: Technological advancements in waste sorting and processing ropeway for SDG 13 (climate action). In R. Mehra, P. Kumar, & B.

Hung (Eds.), *Municipal Solid Waste Management and Recycling Technologies* (pp. 343–366). IGI Global. https://doi.org/10.4018/979-8-3693-4054-7.ch013 (accessed on 5th January 2026).

12. Singh, B., Kaunert, C., & Jermsittiparsert, K. (2024). Environment-biodiversity protection and SDG 15 (life on land): Legal landscapes and probable resolutions. In B. Singh, C. Kaunert, K. Vig, & S. Dutta (Eds.), *Maintaining a Sustainable World in The Nexus of Environmental Science and AI* (pp. 259–282). IGI Global. https://doi.org/10.4018/979-8-3693-6336-2.ch010 (accessed on 5th January 2026).
13. Brookes, G., & Smyth, S. J. (2024). Risk-appropriate regulations for gene-editing technologies. *GM Crops & Food, 15*(1), 1–4.
14. Rueda, J. (2024). The global governance of genetic enhancement technologies: Justification, proposals, and challenges.
15. Adeel, M., & Jones, M. G. (2024). Challenges and prospects in the regulation of CRISPR-edited crops. *Global Regulatory Outlook for CRISPRized Plants*, 447–459.
16. Singh, B., & Kaunert, C. (2024). Transforming smart city and smart sustainable tourism projecting artificial intelligence and IoT: Immersive potential for revolutionizing urban living and enhancing travel experience. In K. Jermsittiparsert & P. Suanpang (Eds.), *Technological and Managerial Approaches to Fostering Sustainable Travel* (pp. 55–80). IGI Global. https://doi.org/10.4018/979-8-3693-8115-1.ch003 (accessed on 5th January 2026).
17. Raghav, A., Singh, B., Jermsittiparsert, K., Raghav, R., & Yadav, U. (2024). Artificial intelligence in environmental and climate changes: A sustainable future. In B. Singh, C. Kaunert, K. Vig, & S. Dutta (Eds.), *Maintaining a Sustainable World in the Nexus of Environmental Science and AI* (pp. 485–506). IGI Global. https://doi.org/10.4018/979-8-3693-6336-2.ch019 (accessed on 5th January 2026).
18. Kaushik, T. K., Ravish, Singh, A., Raghav, A., Singh, B., Arora, M. K., & Lal, S. (2024). Untapping artificial intelligence in environmental and sustainable progression: Projecting legal and ethical considerations. In B. Singh, C. Kaunert, K. Vig, & S. Dutta (Eds.), *Maintaining a Sustainable World in the Nexus of Environmental Science and AI* (pp. 109–130). IGI Global. https://doi.org/10.4018/979-8-3693-6336-2.ch005 (accessed on 5th January 2026).
19. Singh, B., Kaunert, C., Malviya, R., Lal, S., & Arora, M. K. (2024). Scrutinizing consumer sentiment on social media and data-driven decisions for business insights: Fusion of artificial intelligence (AI) and business intelligence (BI) foster sustainable growth. In A. Natarajan, M. Galety, C. Iwendi, D. Das, & A. Shankar (Eds.), *Intersection of AI and Business Intelligence in Data-Driven Decision-Making* (pp. 183–210). IGI Global. https://doi.org/10.4018/979-8-3693-5288-5.ch007 (accessed on 5th January 2026).
20. Barlevy, D., Juengst, E., Kahn, J., Moreno, J., Lambert, L., Charo, A., Chneiweiss, H., Farooque, M., Guston, D. H., Hyun, I., & Knoepfler, P. S. (2024). Governing with public engagement: An anticipatory approach to human genome editing. *Science and Public Policy*, scae010.
21. Karunarathna, I., Hapuarachchi, T., Ekanayake, U., & Gunathilake, S. (2024). The ethics of genetic editing: Navigating the future of science.

22. Sagawa, C. H., Assis, R. D., & Zaini, P. A. (2024). Regulatory framework of CRISPR-edited crops in the United States. In *Global Regulatory Outlook for CRISPRized Plants* (pp. 167–195). Academic Press.
23. Goberna, M. F., Lewi, D. M., Godoy, P., Simeone, F., & Hopp, H. E. (2024). Regulatory framework for CRISPR-edited crops in Argentina. In *Global Regulatory Outlook for CRISPRized Plants* (pp. 213–230). Academic Press.
24. Arif, A., Jamil, A., Munir, A., Ahmad, A., & Munawar, N. (2024). Regulatory, ethical, and social aspects of CRISPR crops. In *CRISPRized Horticulture Crops* (pp. 439–470). Academic Press.
25. Munawar, N., Faheem, M., Niamat, A., Munir, A., Khan, S. H., Zahoor, M. K., Aslam, A., & Ahmad, A. (2024). Regulatory, ethical, social, and biosafety concerns in genome-edited horticultural crops. In *CRISPRized Horticulture Crops* (pp. 421–438). Academic Press.
26. Clark, L. F., & Hobbs, J. E. (2024). What's next for gene editing in agrifood? In *International Regulation of Gene Editing Technologies in Crops: Current Status and Future Trends* (pp. 95–103). Cham: Springer Nature Switzerland.
27. Geuverink, W. P., Houtman, D., Retel Helmrich, I. R., van Baalen, S., van Beers, B. C., van El, C. G., Henneman, L., Kasprzak, M. D., Arets, D., Riedijk, S. R., & 'De DNA dialogen' consortium. (2024). The need to set explicit goals for human germline gene editing public dialogues. *Journal of Community Genetics*, *8*, 1–7.
28. Puchta, H. (2024). Regulation of gene-edited plants in Europe: From the valley of tears into the shining sun? *Abiotech, 5*(2), 231–238.
29. Fernandes, P. M., Favaratto, L., Merchán-Gaitán, J. B., Pagliarini, R. F., Zerbini, F. M., & Nepomuceno, A. L. (2024). Regulation of CRISPR-edited plants in Latin America. In *Global Regulatory Outlook for CRISPRized Plants* (pp. 197–212). Academic Press.
30. Soni, S. (2024). Recent developments in the regulation of heritable human genome editing. *Journal of Bioethical Inquiry, 21*(1), 15–18.
31. Selin, C., Friedman, L. M., Lambert, L., Farooque, M., & Guston, D. (2024). Inclusive, equitable and just evolution of gene editing technologies: The case for integrating public engagement in governing efforts. https://papers.ssrn.com/sol3/papers.cfm?abstract_id=4853681 (accessed on 5th January 2026).
32. Chiong, S. B., & Austriaco, N. (2024). Germline gene editing with CRISPR: A risk-analysis response to liberal eugenics. In *The Routledge Handbook of the Ethics of Human Enhancement* (pp. 154–161). Routledge.
33. Kumawat, T., Agarwal, A., Saxena, S., & Arora, S. (2024). An analysis of global policies and regulation on genome editing in plants. In *Gene Editing in Plants: CRISPR-Cas and its Applications* (pp. 775–793).
34. Abd-Elsalam, K. A., & Ahmad, A. (2024). Regulations of genome edited plants: Notes from editors. In *Global Regulatory Outlook for CRISPRized Plants* (pp. 1–20). Academic Press.
35. Ghag, S. B. (2024). Genetically modified organisms and their regulatory frameworks. In *Global Regulatory Outlook for CRISPRized Plants* (pp. 147–166). Academic Press.
36. Snuzik, A. (2024). Assessing CRISPR/Cas9 potential in SDG3 attainment: Malaria elimination—regulatory and community engagement landscape. *Malaria Journal, 23*(1), 192.

37. Roberts, A. J., Hackett, K., Coche, I., James, S. L., Littler, K., Santos, M., & Emerson, C. I. (2024). Taking stock: Is gene drive research delivering on its principles? *Gates Open Research, 8*(14), 14.
38. Henderson, K., Lang, B., Kemper, J., & Conroy, D. (2024). Exploring diverse food system actor perspectives on gene editing: A systematic review of socio-cultural factors influencing acceptability. *Agriculture and Human Values, 41*(2), 883–907.
39. Abkallo, H. M., Arbuthnot, P., Auer, T. O., Berger, D. K., Burger, J., Chakauya, E., Concordet, J. P., Diabate, A., Di Donato, V., Groenewald, J. H., & Guindo, A. (2024). Making genome editing a success story in Africa. *Nature Biotechnology, 42*(4), 551–554.
40. Li, T. (2024). A Chinese model for legal regulation of gene-edited endangered animals and plants. *Gene, 148350.*
41. Ahmad, A., Munir, A., Munawar, N., Khan, S. H., Khan, Z., & Akhtar, B. (2024). Regulations of multiplex genome-edited crops and CRISPR/Cas gene drives. In *Global Regulatory Outlook for CRISPRized Plants* (pp. 397–445). Academic Press.
42. Singh, B., Kaunert, C., Jermsittiparsert, K., Lal, S., Arora, M. K., & Raghav, A. (2024). Ultimate deep learning acclimatization with artificial intelligence drones for environment shield and climate change: Lifting sustainability towards fostering SDG 13. In B. Singh, C. Kaunert, K. Vig, & S. Dutta (Eds.), *Maintaining a Sustainable World in the Nexus of Environmental Science and AI* (pp. 59–80). IGI Global. https://doi.org/10.4018/979-8-3693-6336-2.ch003 (accessed on 5th January 2026).
43. Singh, B., Kaunert, C., & Gautam, R. (2024). Harnessing machine learning intelligence against cyber threats: Navigating legal and ethical challenges in global cyber investigations. In F. Özsungur (Ed.), *Empowering Entrepreneurial Mindsets with AI* (pp. 319–352). IGI Global. https://doi.org/10.4018/979-8-3693-7658-4.ch016 (accessed on 5th January 2026).
44. Singh, B., & Kaunert, C. (2024). AI-driven strategies for customer engagement, market segmentation, and resource optimization: Projecting end user satisfaction and futuristic growth of business. In V. Nadda, P. Tyagi, A. Singh, & V. Singh (Eds.), *AI Innovations in Service and Tourism Marketing* (pp. 104–128). IGI Global. https://doi.org/10.4018/979-8-3693-7909-7.ch006 (accessed on 5th January 2026).
45. Singh, B., & Kaunert, C. (2025). Dynamic landscape of artificial general intelligence (AGI) for advancing renewable energy in urban environments: Synergies with SDG 11—sustainable cities and communities lensing policy and governance. In S. El Hajjami, K. Kaushik, & I. U. Khan (Eds.), *Artificial General Intelligence (AGI) Security*. Singapore: Springer. https://doi.org/10.1007/978-981-97-3222-7_12 (accessed on 5th January 2026).
46. Singh, B., & Kaunert, C. (2024). Blue ocean and machine learning trajectories SDG 14: Life below water for handling ocean pollution: Metaverse conserve ocean health sustainability through the lens of transboundary legal-policy regulations as articulating space for futuristic changes. In D. De, D. Sengupta, & T. A. Tran (Eds.), *Artificial Intelligence and Edge Computing for Sustainable Ocean Health*. Cham: Springer. https://doi.org/10.1007/978-3-031-64642-3_4 (accessed on 5th January 2026).
47. Dutta, P. K., Singh, B., Towfeek, A. S., Adamopoulou, J. P., Bardavouras, A. N., Bamwerinde, W., et al. (2024). IoT revolutionizes humidity measurement and

management in smart cities to enhance health and wellness. *Mesopotamian Journal of Artificial Intelligence in Healthcare*, 110–117. https://doi.org/10.58496/MJAIH/2024/013 (accessed on 5th January 2026).

48. Irfan, M., Majeed, H., Iftikhar, T., & Ravi, P. K. (2024). A review on molecular scissoring with CRISPR/Cas9 genome editing technology. *Toxicology Research, 13*(4).
49. Burgess, J. (2024). Balancing progress and ethics: Exploring the science and ethics of gene editing: Literature review. https://scholarworks.uni.edu/hpt/942/ (accessed on 5th January 2026).
50. Aggarwal, A., Komal, D. S., & Chaudhary, R. (2024). A critical analysis of legal boundaries in biotechnology and biosecurity. *MSW Management Journal, 34*(1), 239–249.
51. Wu, Y. (2024). The development of gene editing technology and controversial issues: A discussion. *Highlights in Science, Engineering and Technology, 91*, 123–130.
52. Singh, B., & Kaunert, C. (2024). Hitching artificial intelligence and IoT and for livestock management competence: Line up nature and technology for smart agriculture. In M. Zia Ul Haq & I. Ali (Eds.), *Revolutionizing Pest Management for Sustainable Agriculture* (pp. 499–520). IGI Global. https://doi.org/10.4018/979-8-3693-3061-6.ch020 (accessed on 5th January 2026).
53. Singh, B., & Kaunert, C. (2024). Tagging blockchain technology fostering panacea for data privacy, cloud computing, and integrity: Legal framework in India and at a globe. In *Artificial Intelligence Solutions for Cyber-Physical Systems* (pp. 206–213). Auerbach Publications. https://www.taylorfrancis.com/chapters/edit/10.1201/9781032694375-11/tagging-blockchain-technology-fostering-panacea-data-privacy-cloud-computing-integrity-bhupinder-singh-christian-kaunert (accessed on 5th January 2026).
54. Singh, B., Raghav, A., & Jermsittiparsert, K. (2024). Potential benefits of artificial intelligence and machine learning to overcome health consequences of sexual violence: The case of Thailand and India. *Asian Crime and Society Review, 11*(2), 102–108.
55. Singh, B., & Kaunert, C. (2024). Redefining the nodes of metaverse and luxury hospitality: Unveiling AI and tech-driven solutions in hospitality and tourism industry. In M. Talukder, S. Kumar, & P. Tyagi (Eds.), *Impact of AI and Tech-Driven Solutions in Hospitality and Tourism* (pp. 175–202). IGI Global. https://doi.org/10.4018/979-8-3693-6755-1.ch010 (accessed on 5th January 2026).
56. Shen, R., Shen, X., & Wu, P. (2024). Economic impact of gene editing technology: The contribution of personalized medicine, bioinformatics, and artificial intelligence. *Journal of Simulation, 12*(1), 1.
57. Lassoued, R., Smyth, S. J., & Phillips, P. W. (2024). Regulatory outlook of CRISPR-edited plants in Canada. In *Global Regulatory Outlook for CRISPRized Plants* (pp. 231–250). Academic Press.
58. Ahmad, A., Arif, A., & Munir, A. (2024). CRISPR technology commercialization and biosafety. In *Global Regulatory Outlook for CRISPRized Plants* (pp. 461–514). Academic Press.
59. Singh, B., & Kaunert, C. (2024). Trans-disciplinary collaborative disciplines generating cohesive curriculum: Runway for innovation, cooperation, and

intentionality in higher education. In R. Kumar, E. Ong, S. Anggoro, T. Toh, & M. Fukui (Eds.), *Transdisciplinary Teaching and Technological Integration for Improved Learning: Case Studies and Practical Approaches* (pp. 152–176). IGI Global. https://doi.org/10.4018/979-8-3693-8217-2.ch008 (accessed on 5th January 2026).

60. Kumar, S., Malviya, R., Uniyal, P., Srivastava, S., Sridhar, S. B., Obaidur Rab, S., et al. (2024). Plant fiber composites for biomedical applications: Advances and prospective. *Current Pharmaceutical Biotechnology*. https://doi.org/10.2174/0113892010314833240815113622 (accessed on 5th January 2026).
61. Singh, B., & Kaunert, C. (2024). Marvel of biosensors in smart healthcare and application of Internet of Medical Things for diagnosis: Unveiling benefits, challenges and futuristic approach. In *Smart Healthcare Systems* (pp. 187–198). CRC Press. https://www.taylorfrancis.com/chapters/edit/10.1201/9781032698519-14/marvel-biosensors-smart-healthcare-application-internet-medical-things-diagnosis-bhupinder-singh-christian-kaunert (accessed on 5th January 2026).
62. Singh, B., Kaunert, C., & Jermsittiparsert, K. (2025). Advocating medical tourism and healthcare in global health marketplaces: Promoting sustainability and technological advancements. In K. Jermsittiparsert & R. Marzo (Eds.), *AI Technologies and Advancements for Psychological Well-Being and Healthcare* (pp. 327–350). IGI Global. https://doi.org/10.4018/979-8-3693-9158-7.ch012 (accessed on 5th January 2026).
63. Singh, B., & Kaunert, C. (2025). Blockchain framework for precision medicine, clinical trials, and genomic biomarkers discovery: Focuses on monitoring and managing data in health dimensions. In K. Jermsittiparsert & R. Marzo (Eds.), *AI Technologies and Advancements for Psychological Well-Being and Healthcare* (pp. 255–288). IGI Global. https://doi.org/10.4018/979-8-3693-9158-7.ch010 (accessed on 5th January 2026).
64. Singh, N., Lal, S., Raghav, A., Arora, M. K., Singh, B., & Kaunert, C. (2024). Power of ML and DL in precision modeling and climate predictions: Next-gen ocean insights. In B. Riswandi, B. Singh, C. Kaunert, & K. Vig (Eds.), *AI Applications for Clean Energy and Sustainability* (pp. 254–273). IGI Global. https://doi.org/10.4018/979-8-3693-6567-0.ch013 (accessed on 5th January 2026).
65. Singh, B., Kaunert, C., & Kaushik, T. K. (2024). Unscrambling financial fraud with AI and machine learning in e-commerce transactions: Airing into ad clicks, credit card management. In B. Pandow, F. Masoodi, J. Iqbal, & G. Hussain (Eds.), *Navigating the Future of Finance in the Age of AI* (pp. 253–271). IGI Global. https://doi.org/10.4018/979-8-3693-4382-1.ch013 (accessed on 5th January 2026).
66. Nelson, J. P., Tomblin, D. C., Barbera, A., & Smallwood, M. (2024). The divide so wide: Public perspectives on the role of human genome editing in the US healthcare system. *Public Understanding of Science, 33*(2), 189–209.
67. Singh, B., Kaunert, C., Lal, S., Arora, M. K., & Singh, G. (2024). Abundance ocean wave energy to electricity with artificial intelligence and IoT solutions: Future of clean energy and sustainability in smart cities. In B. Riswandi, B. Singh, C. Kaunert, & K. Vig (Eds.), *AI Applications for Clean Energy and Sustainability* (pp. 274–298). IGI Global. https://doi.org/10.4018/979-8 (accessed on 5th January 2026).

CHAPTER 2

Application and Challenges of CRISPR/Cas9 System for Gene Editing

THANAPORN SRIYAKUL,[1] SAQUIB AHMED,[2] BHUPINDER SINGH,[2,3*] and ANJALI RAGHAV[2]

[1]*Association of Legal and Political Studies, Bangkok, Thailand*

[2]*Sharda School of Law, Sharda University, Greater Noida, Uttar Pradesh, India*

[3]*University of South Wales, UK; Santo Tomas University Tunja, Colombia*

*Corresponding author

ABSTRACT

The CRISPR/Cas system is a recently developed and promising method for gene editing. This simple approach, which has great therapeutic promise, was developed based on the prokaryotic adaptive immune defense mechanism and used in studies on human diseases. The CRISPR technique can be used to repair a genetically distinct patient mutation during gene therapy, treating diseases that conventional approaches were unable to treat. The introduction of CRISPR/Cas9 to the healthcare environment will be difficult because there is still more, we can do to enhance the technology's applications, precision, and efficacy. In this chapter, we will first describe the CRISPR/Cas9 system (hereinafter referred to as the System) and then its function and application. It outlines the prospective uses of this technique for gene therapy of different human problems, such as infectious diseases and cancer, and we emphasize the promising instances in the field. Lastly, we list the challenges that still need to be resolved to effectively apply CRISPR-Cas9 in healthcare.

Genetic Editing for Healthcare: Advances in Disease Management.
Bhupinder Singh, Rishabha Malviya, Sahil Lal, and Sathvik Belagodu Sridhar (Eds.)

DOI: 10.1201/9781779643827-2

2.1 INTRODUCTION

Gene editing technology is a type of genome engineering that involves the alteration and modification of DNA sequences or their variations, such as addition, deletion, integration, and replacement, using a sequence-based approach [1]. Genome-editing methods: Editing within the genome is performed in a way that cellular DNA repair mechanisms have been derived. Directed genetic and chemical (epigenetic) changes by enzyme-based tools can be utilized to modify the genome in a site-specific manner [2]. Despite its recent history, CRISPR has been repurposed quickly to work as a potent genome editing system [3] that provides immunity against invading viruses in many prokaryotes and archaea (including humans). CRISPR has naturally evolved as an immune system in many organisms. CRISPR-associated (Cas) enzymes cleave specific nucleic acid sequences using the CRISPR technology [4]. The ability to program CRISPRs with high reliability is making gene-editing techniques faster and more effective than traditional methods for therapeutic genetic editing, which are largely being replaced by this powerful technology as a "go-to" tool [5]. The system has shown promise as a way to correct gene edits in diseases such as blood disorders [6], degenerative muscle disease [7], and neurological, cardiovascular, and renal genetic-based conditions of the genes, stem cells, and optics-related diseases [8, 9]. When the genetic cause of a malfunction is identified, the system offers remarkable therapeutic promise for treating a variety of diseases [10]. By creating many disease models, the system advances our understanding of these disorders [11]. Reprogramming therapies based on CRISPR-Cas9-mediated genome editing may restore gene function or make up for a mutation [12].

There are numerous CRISPR-based gene-editing systems known to exist, and each has unique needs for nucleic acid binding and enzymatic activity [6]. Classifying a number of CRISPR-Cas systems is crucial for improving our comprehension of their origins and facilitating future research [13]. Differences in the "Cas protein" compositions and the sequence variations between effector complexes are the basis for categorizing CRISPR systems [14].

2.2 DEVELOPMENT AND LIMITATIONS OF CRISPR GENOME EDITING

CRISPR (clustered regularly interspaced short palindromic repeats) genome editing is one of the most significant turning points in modern science, which changed our approach to genetic research and therapeutic strategies. CRISPR's genome-editing journey was rooted in the discovery of a bacterial immune system. CRISPR is part of the bacterial defense system that chops up foreign viral DNA. This mechanism could be repurposed as a tool for editing the genomes of eukaryotic cells, marking one of the most important technological advances in science today, which allows researchers to make specific changes at any location within an organism's genome.

The CRISPR-Cas9 system functions by directing a protein called Cas9 to an exact sequence of DNA with the aid of RNAs that can be specifically designed. Using the Cas9 protein, which functions as molecular scissors that slice DNA at a specific spot, this cut makes the cell do what it does naturally: heal (disabling a gene, inactivating it, or adding another piece of genetic material). CRISPR provides the unique ability to target and edit genetic sequences with unprecedented precision, so it has enormous value as a research tool as well as in clinical contexts.

The discovery of CRISPR technology paved the way for new horizons in medicine, agriculture, and environmental conservation. CRISPR Could Cure All Diseases. In medicine, CRISPR technology may correct an array of genetic disorders such as cystic...medium.com. Within agriculture, CRISPR technology has been deployed to create crops with improved resistance toward pests and diseases, as well as abiotic stresses that can increase yields or improve food security. Beyond that, CRISPR is used in environmental science to combat invasive species and engineer organisms capable of degrading the environment.

CRISPR genome editing, while promising to revolutionize fields ranging from agriculture to medicine, is limited in many respects. A key concern is what are called "off-target effects," when the CRISPR-Cas9 system mistakenly snips DNA in places other than where it was designed to, creating unwanted mutations. Such off-target edits might have unexpected health consequences if CRISPR is ever deployed in therapeutic settings. Scientists are currently developing ways to increase the specificity of the CRISPR system to reduce these off-target effects, but it is still a problem that must be solved before this sees use in humans.

Immune rejection against CRISPR components is another limitation. Because the human body recognizes Cas9 as a bacterial protein, it can trigger an immune response that might limit how well CRISPR can work when used therapeutically or could harm people. Despite ongoing work, the use of such systems for therapeutic purposes is still hampered by host immune responses, and new or modified CRISPR approaches that avoid detection are likely to be required. There is another very important limit to the technique: it's completely unethical. With great power comes not just responsibility but also a plethora of ethical questions, most notably around germline editing, as the genetic changes are inheritable by those yet to come. The ability to change the human germline could be used to eliminate inherited genetic diseases, which should help most critics support further research; however, it will also enable controversial practices such as naturally designed babies ultimately picked for enrichment traits instead of health conditions. Regulations in other countries expressly forbid germline modification, and international guidelines have counseled the scientific community to refrain from using CRISPR technologies on human embryos.

With the extensive and diverse applications of CRISPR genome editing in bench research to clinical practice, significant achievements have been made, but challenges are still faced during its development: technical hurdles for efficiency (e.g., on- or off-target efficiency) as well as ethical controversies (biosecurity risks including human consumption safety, regulatory concerns) that warrant more investigation. Further research is needed to increase the precision and safety of this technique, where ethical considerations regarding its use exist, especially when it concerns human genetic information.

2.2.1 HISTORICAL BACKGROUND

Initially, advances in eukaryotic DNA repair gave rise to genome editing. Groundbreaking research conducted in the early 1990s, employing homing endonucleases like I-SceI, [15] which recognizes 18-bp DNA sequences [16], showed that the production of a targeted double-strand break (DSB) in mammalian cells enhanced homologous recombination at the target location. This made genome editing with DSB-generating nucleases a reality. The subsequent development of engineered nuclease enzymes, based on complex fusions of nonspecific DNA endonucleases with tandem arrays of sequence-specific DNA binding modules, was prompted by the need for diverse nucleases with long recognition sites that could target single sites in eukaryotic genomes [17]. The early 2000s saw the introduction of zinc-finger nucleases (ZFNs), which were followed in 2010–2011 by transcription activator-like effector nucleases (TALENs). The discovery and development of RNA-guided [18], CRISPR-associated (Cas) nucleases brought about a revolutionary change because of their adaptability, specificity, and ease of programming [19].

The adaptive genome defense mechanisms of prokaryotic CRISPR-Cas systems were demonstrated in 2007 Tertulino et al. [20] to identify and target foreign nucleic acids linked to viruses and other mobile genetic elements. Within these systems, CRISPR RNAs (crRNAs) are produced by transcription and subsequent processing of invading DNA fragments that are obtained and stored in repeating arrays [21]. As molecular guides, crRNAs instruct Cas protein-based molecular machinery to identify invasive nucleic acids and direct their destruction [22].

Later studies showed that the so-called type II CRISPR-Cas systems site-specifically cleaved phage DNA, that the activity was dependent on the protein Cas9 [23], and that trans-activating crRNA (tracrRNA) was a necessary RNA component for crRNA maturation [24]. Lastly, biochemical studies published in 2012 showed that Cas9 was a DNA-cleaving endonuclease whose specificity was determined by a dual-RNA guide structure made up of crRNA and tracrRNA [25]. The system was further simplified by combining tracrRNA and crRNA into a single guide RNA (sgRNA) [26], creating a fully programmable one-nuclease-one-guide-RNA design [27].

This served as the foundation for the creation of CRISPR genome editing technologies, which allowed for the creation of sgRNAs with matching sequences to direct Cas9 to a particular genomic location [28]. A race to repurpose Cas9 for genome editing began when its RNA-guided DNA cutting activity was discovered [29]. This race culminated in a number of studies published in early 2013, which reported that the introduction of genetic modification at target genomic loci was caused by the expression of Cas9 and particular sgRNAs in eukaryotic cells [30].

2.3 CLINICAL IMPLICATIONS OF THE CRISPR SYSTEM

2.3.1 CANCER TREATMENT

Mutations in oncogenes and tumor suppressor genes, both genetic and epigenetic, result in cancer, a complicated illness [31]. In tumor research, CRISPR-Cas9 can be used to alter the genome and look into the causes of tumor initiation, growth, and metastasis [32]. In recent years, this technique has been extensively used in research on cancer treatments by deleting certain genes or fixing mutations [33]. So far, a large number of exploratory studies on tumor treatment in related fields have been conducted [34]. Presently, cancer, viral infections, and single-gene diseases are the main targets of CRISPR-Cas9 [35]. Currently, blood malignancies including lymphoma and leukemia are treated with CRISPR-based medicines [36]. Furthermore, a lung cancer trial that was conducted in China has just ended [37]. Treatments for solid tumors [38], gastric carcinoma [39], nasopharyngeal carcinoma [40], hepatocellular carcinoma, renal cell carcinoma, B cell leukemia, and lymphoma are being sought after by a number of studies [41].

2.3.2 INFECTIOUS DISEASE

Nowadays, one of the most prevalent worldwide causes of severe and fatal human illnesses is infectious infections [42]. These illnesses can be challenging to cure, can spark epidemics, and necessitate preventive care. Moreover, a large number of them carry mutations that allow vaccines to evade immunity or be resistant to antibiotics; in this situation, latent infections are expected [43]. Infectious disease can be classified broadly into two groups as: Bacterial Infectious Disease and Viral Infectious Disease.

The cornerstone of treatment for bacterial infections is antibiotic-based drugs; however, the risk of multidrug-resistant (MDR) bacteria is growing significantly as a result of improper and excessive use of these drugs as well as their presence in feedstock, which enables bacteria to constantly evolve and adapt against the drugs [44]. Also, it is simple for one host to transfer the established resistance to another. Nevertheless, there is hope that CRISPR-Cas9, which has emerged as a significant technology that targets the genome of resistant bacteria in a sequence-specific manner, will help better handle this danger [45].

2.3.3 COVID-19

Severe Acute Respiratory Syndrome Coronavirus-2, or SARS-CoV-2, is the source of the recent COVID-19 pandemic [46]. Over the past three years, the virus has killed millions of people globally and severely restricted social and economic opportunities [47]. Since COVID-19 first emerged, the virus has been extensively studied in clinical trials, leading to the development of multiple vaccinations. Many clinical trials are now registered under the COVID-19 keyword in various databases, while some are also connected to CRISPR technology [48]. The trials can be categorized into two groups: *In vitro* diagnostics and *Ex vivo* cell treatments.

In addition to other diagnostic techniques like RTPCR, CRISPR's high specificity and sensitivity led to its quick adoption in diagnostics as a test at the point of care [49]. Early detection and/or treatment of COVID-19 and other related infectious disorders is essential due to their high transmissibility, severe pathogenicity, and propensity for the development of different mutant strains [50]. As a result, CRISPR technology, which can be used in point-of-care testing and enhance immune system function by focusing on viral RNA [51], has a lot of potential for use in pandemics like this as one of the initial choices during the diagnosis and treatment phases [52].

2.3.4 HUMAN IMMUNODEFICIENCY VIRUS (HIV)

HIV-AIDS is an example of one deadly illness that has plagued millions. Despite lifelong antiretroviral medication being able to "shock and kill" Callahan et al. [52] replicating virus, the persistence of HIV provirus [53], a type of latent infection [54], remains one challenge in an eventual cure for HIV. The immune system is unable to detect the infection as HIV suppresses viral protein expression during latency [55]. Thus, HIV DNA interrupts the most challenging part of achieving a cure for people infected with symptomatic providers [56]. The provirus and host genes that are essential for the virus entry have good potential targets for gene therapy to treat HIV [57]. These strategies are now performed using CRISPR/Cas techniques [58].

2.3.5 PERIODONTIC DISORDERS

Periodontitis, a dangerous gum disease that affects the soft connective tissue and causes inflammation and infection, can harm bone tissue [59]. Microbes linked to oral disorders can be targeted with CRISPR technology [60]. For instance, oral infections and biofilms can be targeted by bacteriophage engineering using the System [61]. The CRISPRa, CRISPRi, and Cas13 systems also modify the transcriptome without affecting DNA sequences, which eliminates the genes linked to periodontal disease [62]. This technique has an advantage over conventional DNA editing in that the modifications are transient and reversible [63]. Although there hasn't been much use of the CRISPR technique in periodontitis experimental trials overall, as this system advances, additional therapeutic approaches will take the place of surgery to treat periodontal disease [64].

2.4 CHALLENGES WITH THE CRISPR SYSTEM FOR GENE EDITING

Despite CRISPR/Cas9 being a promising technique, its use in clinical research is challenging due to its recent discovery and use in humans [65]. Immunogenicity, off-targeting, mutations, delivery methods, and ethics are the primary challenges associated with CRISPR/Cas9 technology [66].

2.4.1 IMMUNOGENICITY

The process of a material inciting an immune response leading to severe side effects in the body due to being seen as foreign is called immunogenicity, or immunoreactivity [67]. Researchers have demonstrated that pre-existing immunity can be elicited by exposure to similar epitopes of CRISPR system components through previous antigens [68]; these agents include Cas9, delivery vectors, and guide RNAs. However, this could potentially lead to failed attempts at genome editing [69]. The CRISPR systems have been described to be the source of immune responses in healthy individuals, both innate (by guide RNAs) and acquired anti-Cas9 T-cell cellular [70], including also humoral response. Some strategies to tackle this problem have been proposed, including phosphatase treatment of *in vitro* transcribed guide RNAs [71], new Cas proteins and delivery vectors likely never-before-seen by humans these days being developed for application as CRISPR agents [72], and characterizing host immune responses against components of a recorded prior response following successful therapeutic implementation both preclinical screenings [73].

2.4.2 OFF-TARGETING

One of the main issues with CRISPR/Cas9 is off-targeting. Due to similarities in the genome, the gRNA may bind to non-specific regions when the CRISPR–Cas9 system is used in complex genomic species, like mammals, which could result in further changes [74]. Technologies in bioinformatics have been developed to predict and reduce off-target changes [75]. These could be improved to assist scientists in developing new medications [76].

It is also possible to decrease off-target functions and boost efficiency by using the High-fidelity SpCas9. Longer gRNA frequently offers a better affinity for target genes in HDR-based repair [77]. To lessen the chance of off-target effects, enzymes (nucleases) can be altered, or Cas9 can be temporarily produced [78]. Additionally, the type of vectors can reduce the off-target risk; for example, using adenovirus (AV) vectors prevents incorporation into the host DNA [79].

2.4.3 *MUTATIONS*

Contrary to other genetic conditions like Huntington's disease or hemophilia, cancer is associated with a number of gene mutations [80]. Altering the abnormal nucleotides of tumor suppressor genes is one tactic that should be taken into consideration in order to knock in the mutant gene and accomplish the intended result in cancer therapy [81]. The CRISPR/Cas9 technology can reduce or limit carcinogenesis by modifying the activation of certain tumor suppressor genes [3]. However, changing a single mutated nucleotide is rarely sufficient in cancer gene therapy [82]. In CRISPR-Cas9, correcting faulty nucleotides by knocking in is more challenging than knocking out [83], and as was previously said, numerous gene mutations cause cancer [84].

As a result, knocking in every gene that causes cancer is time-consuming, complicated, and more likely to mistarget. It also requires multiple guide RNAs [85]. To get around the challenges, we can employ the CHyMErA techniques [85], which rely on the Cas9 and Cas12a nucleases and two Cas proteins [86], as well as bioinformatic tools to identify protein interaction sites [87] and forecast the results of CRISPR knock-out [88].

2.4.4 *ETHICS*

Novel insights into human genetic alterations have been provided by CRISPR/Cas9-mediated genome editing [89]. The sociological and ethical implications of this technology must be carefully considered due to its unpredictable nature and wide-ranging effects [90].

2.5 CONCLUSION AND FUTURE DIRECTIONS

Current significant advancements in gene-editing technology are already significantly enhancing human health. For example, gene editors are already being used in clinical trials to treat a variety of human diseases, such as HIV, cancer, and hematologic disorders. New therapies for additional illnesses will probably become available as gene editors advance. In particular, CRISPR systems, which are contemporary gene editors that are developing quickly, have been utilized to cause many genetic alterations in "human cells," including site-specific editing of DNA sequences, activation or suppression of desired gene expression, and epigenetic modifications. To guarantee the safety and efficacy of treatment options for impacted persons, a deeper comprehension of the fundamental biology underlying these gene-editing technologies is required, notwithstanding the possible drawbacks of gene editors for human therapy. Because of the intricate interactions between a tool's molecular function and delivery techniques, the efficacy and safety of CRISPR systems *in vivo* remain uncertain. Some CRISPR ingredients have caused immunogenic responses in humans, and in many cases, using CRISPR-based gene-editing tools has widespread off-target effects, which can be dangerous if we have to use them in a vital organ or if they are accidentally administered.

It is important to look at new ways to improve distribution to target sites and decrease off-target sites. Research teams are currently thoroughly examining these problems, and developments in these fields will be crucial to the success of gene-editing treatments. The quick development of new gene-editing technologies in recent years has also raised a number of ethical concerns, ranging from modifying a human embryo in development to deciding who should have the power to enact laws or make decisions. To regulate and supervise the use of CRISPR and genome editing technologies in developing children, there will need to be an international discussion between science and politics. Decisions must be based on a known or possible compromise since, contrary to popular opinion, health and sickness frequently contain a range. Furthermore, there are significant hazards associated with genome editing; consequently, in each case, a balance between the recognized risks and the significant benefits must be achieved. As technology develops and the risk/benefit ratio for a particular treatment shift, this balance will also change. There is worry that new technologies could be intentionally or accidentally exploited as they advance.

The strong gene-editing technologies, particularly CRISPR, have promise for developing novel treatments for a wide range of human illnesses. Their clinical growth is almost a reality since more and more money is being spent annually to better comprehend and describe these technologies. Last but not least, CRISPR/Cas9 cell-based regeneration therapies have the potential to solve the donor compatibility issues related to transplant rejection. To fix a mutation in a patient's own tissues, these processes, known as autologous therapies, invoke genetic modification.

KEYWORDS

- **COVID-19**
- **CRISPR/Cas**
- **gene editing**
- **immunogenicity**
- **mutations**
- **periodontitis**
- **SARS-CoV-2**

REFERENCES

1. Gratz, S., O'Connor-Giles, K. M., & Wildonger, J. (2024). Introduction to CRISPR and its use in *Drosophila. Cold Spring Harbor Protocols, 2024*(5), pdb-top108228.
2. Morshedzadeh, F., Ghanei, M., Lotfi, M., Ghasemi, M., Ahmadi, M., Najari-Hanjani, P., Sharif, S., Mozaffari-Jovin, S., Peymani, M., & Abbaszadegan, M. R. (2024). An update on the application of CRISPR technology in clinical practice. *Molecular Biotechnology, 66*(2), 179–197.
3. Irfan, M., Majeed, H., Iftikhar, T., & Ravi, P. K. (2024). A review on molecular scissoring with CRISPR/Cas9 genome editing technology. *Toxicology Research, 13*(4), tfae105.
4. Harsij, Z., Ghafoorzadeh, Z., & Goharian, E. (2024). The CRISPR revolution: Unraveling the mysteries of life's genetic code. *Gene, 892*, 147870.
5. Chibuike, M. C., Sara, G. S., & Adele, B. (2024). Overcoming challenges for improved patient-centric care: A scoping review of platform ecosystems in healthcare. *IEEE Access*.
6. Adeniyi, A. O., Arowoogun, J. O., Okolo, C. A., Chidi, R., & Babawarun, O. (2024). Ethical considerations in healthcare IT: A review of data privacy and patient consent issues. *World Journal of Advanced Research and Reviews, 21*(2), 1660–1668.

7. Sharma, P., Namasudra, S., Crespo, R. G., Parra-Fuente, J., & Trivedi, M. C. (2023). EHDHE: Enhancing security of healthcare documents in IoT-enabled digital healthcare ecosystems using blockchain. *Information Sciences, 629*, 703–718.
8. Rai, N., Singh, N. A., Goyal, J., Upadhayay, P., & Jain, V. (2024). CRISPR-Cas and its applications in food production. In *Gene Editing in Plants: CRISPR-Cas and Its Applications* (pp. 349–391). Singapore: Springer Nature Singapore.
9. Li, S., Jin, B., Ma, Y., Yang, X., Fan, J., Xie, Y., Xu, C., Dai, X., Wang, M., Liu, Q., & Fu, T. (2024). Proteome fishing for CRISPR/Cas12a-based orthogonal multiplex aptamer sensing. *Journal of the American Chemical Society, 146*(29), 19874–19885.
10. Gattani, A., Mandal, S., Agrawal, A., Patel, P., Jain, A. K., Singh, P., Garg, A., & Mishra, A. (2024). CRISPR-based electrochemical biosensors for animal health: Recent advances. *Progress in Biophysics and Molecular Biology*.
11. Vasconcelos Komninakis, S., Domingues, W., Saeed Sanabani, S., Angelo Folgosi, V., Neves Barbosa, I., & Casseb, J. (2024). CRISPR/Cas as a powerful tool for human immunodeficiency virus cure: A review. *AIDS Research and Human Retroviruses*.
12. Ruokolainen, J., Nätti, S., Juutinen, M., Puustinen, J., Holm, A., Vehkaoja, A., & Nieminen, H. (2023). Digital healthcare platform ecosystem design: A case study of an ecosystem for Parkinson's disease patients. *Technovation, 120*, 102551.
13. Adeghe, E. P., Okolo, C. A., & Ojeyinka, O. T. (2024). Evaluating the impact of blockchain technology in healthcare data management: A review of security, privacy, and patient outcomes. *Open Access Research Journal of Science and Technology, 10*(2), 013–020.
14. Shah, V., Thakkar, V., & Khang, A. (2023). Electronic health records security and privacy enhancement using blockchain technology. In *Data-Centric AI Solutions and Emerging Technologies in the Healthcare Ecosystem* (pp. 1–13). CRC Press.
15. Lepore, D., Frontoni, E., Micozzi, A., Moccia, S., Romeo, L., & Spigarelli, F. (2023). Uncovering the potential of innovation ecosystems in the healthcare sector after the COVID-19 crisis. *Health Policy, 127*, 80–86.
16. Ahmed, S., & Khan, M. (2023). Securing the Internet of Things (IoT): A comprehensive study on the intersection of cybersecurity, privacy, and connectivity in the IoT ecosystem. *AI, IoT and the Fourth Industrial Revolution Review, 13*(9), 1–7.
17. Aminabee, S. (2024). The future of healthcare and patient-centric care: Digital innovations, trends, and predictions. In *Emerging Technologies for Health Literacy and Medical Practice* (pp. 240–262). IGI Global.
18. Nowrozy, R., Ahmed, K., Kayes, A. S., Wang, H., & McIntosh, T. R. (2024). Privacy preservation of electronic health records in the modern era: A systematic survey. *ACM Computing Surveys, 56*(8), 1–37.
19. Khang, A., Hajimahmud, A. V., Triwiyanto, T., Abuzarova, V. A., & Ali, R. N. (2024). Cloud platform and data storage systems in the healthcare ecosystem. In *Medical Robotics and AI-Assisted Diagnostics for a High-Tech Healthcare Industry* (pp. 343–356). IGI Global.
20. Tertulino, R., Antunes, N., & Morais, H. (2024). Privacy in electronic health records: A systematic mapping study. *Journal of Public Health, 32*(3), 435–454.
21. Pacesa, M., Pelea, O., & Jinek, M. (2024). Past, present, and future of CRISPR genome editing technologies. *Cell, 187*(5), 1076–1100.

22. Li, L., Zhang, D., Zhang, Z., & Zhang, B. (2024). CRISPR/Cas: A powerful tool for designing and improving oil crops. *Trends in Biotechnology*.
23. De Rouck, S., Mocchetti, A., Dermauw, W., & Van Leeuwen, T. (2024). SYNCAS: Efficient CRISPR/Cas9 gene-editing in difficult to transform arthropods. *Insect Biochemistry and Molecular Biology, 165*, 104068.
24. Pandya, K., Jagani, D., & Singh, N. (2024). CRISPR-Cas systems: Programmable nuclease revolutionizing the molecular diagnosis. *Molecular Biotechnology, 66*(8), 1739–1753.
25. Khoshandam, M., Soltaninejad, H., Hamidieh, A. A., & Hosseinkhani, S. (2024). CRISPR, CAR-T, and NK: Current applications and future perspectives. *Genes & Diseases, 11*(4), 101121.
26. Gupte, P., Dhingra, K., & Saloni, V. (2024). Precision gene editing strategies with CRISPR-Cas9 for advancing cancer immunotherapy and Alzheimer's disease. *Journal of Knowledge Learning and Science Technology, 3*(4), 11–21.
27. D'Amato, R., Taxiarchi, C., Galardini, M., Trusso, A., Minuz, R. L., Grilli, S., Somerville, A. G., Shittu, D., Khalil, A. S., Galizi, R., & Crisanti, A. (2024). Anti-CRISPR *Anopheles* mosquitoes inhibit gene drive spread under challenging behavioral conditions in large cages. *Nature Communications, 15*(1), 952.
28. Asmamaw Mengstie, M., Teshome Azezew, M., Asmamaw Dejenie, T., Teshome, A. A., Tadele Admasu, F., Behaile Teklemariam, A., Tilahun Mulu, A., Mekonnen Agidew, M., Adugna, D. G., Geremew, H., & Abebe, E. C. (2024). Recent advancements in reducing the off-target effect of CRISPR-Cas9 genome editing. *Biologics: Targets and Therapy*, 21–28.
29. Oluwaseyi, J. (2024). Genetic engineering and CRISPR: The chemical tools and techniques used in gene editing and their potential applications in medicine.
30. Khare, P., Mandal, S., Verma, A. K., Parashar, D., Jadhav, K., & Gupta, S. (2024). CRISPR: The Janus god of modern science. In *CRISPR-Cas System in Translational Biotechnology* (pp. 23–56). Academic Press.
31. Khoshandam, M., Soltaninejad, H., Mousazadeh, M., Hamidieh, A. A., & Hosseinkhani, S. (2024). Clinical applications of the CRISPR/Cas9 genome-editing system: Delivery options and challenges in precision medicine. *Genes & Diseases, 11*(1), 268–282.
32. Moustafa, K. (2024). CRISPR: Beyond the excitement. *Journal of Bioethical Inquiry*, 1–3.
33. Villiger, L., Joung, J., Koblan, L., Weissman, J., Abudayyeh, O. O., & Gootenberg, J. S. (2024). CRISPR technologies for genome, epigenome and transcriptome editing. *Nature Reviews Molecular Cell Biology*, 1–24.
34. Severi, A. A., & Akbari, B. (2024). CRISPR-Cas9 delivery strategies and applications: Review and update. *Genesis, 62*(3), e23598.
35. Kassam, I., Ilkina, D., Kemp, J., Roble, H., Carter-Langford, A., & Shen, N. (2023). Patient perspectives and preferences for consent in the digital health context: State-of-the-art literature review. *Journal of Medical Internet Research, 25*, e42507.
36. Khang, A., Hahanov, V., Litvinova, E., Chumachenko, S., Hajimahmud, A. V., Ali, R. N., Alyar, A. V., & Anh, P. T. (2023). The analytics of hospitality of hospitals in a healthcare ecosystem. In *Data-Centric AI Solutions and Emerging Technologies in the Healthcare Ecosystem* (pp. 39–61). CRC Press.

37. Fortuna, K. L., Kadakia, A., Cosco, T. D., Rotondi, A., Nicholson, J., Mois, G., Myers, A. L., Hamilton, J., Brewer, L. C., Collins-Pisano, C., & Barr, P. (2023). Guidelines to establish an equitable mobile health ecosystem. *Psychiatric Services, 74*(4), 393–400.
38. Jimo, S., Abdullah, T., & Jamal, A. (2023). IoE security risk analysis in a modern hospital ecosystem. In *Cybersecurity in the Age of Smart Societies: Proceedings of the 14th International Conference on Global Security, Safety and Sustainability, London, September 2022* (pp. 451–467). Cham: Springer International Publishing.
39. Megaro, A., Carrubbo, L., Polese, F., & Sirianni, C. A. (2023). Triggering a patient-driven service innovation to foster the service ecosystem well-being: A case study. *The TQM Journal, 35*(5), 1256–1274.
40. Curtis, C., & Brolan, C. E. (2023). Health care in the metaverse. *Medical Journal of Australia, 218*(1), 46.
41. Nakayama, L. F., de Matos, J. C., Stewart, I. U., Mitchell, W. G., Martinez-Martin, N., Regatieri, C. V., & Celi, L. A. (2023). Retinal scans and data sharing: The privacy and scientific development equilibrium. *Mayo Clinic Proceedings: Digital Health, 1*(2), 67–74.
42. Raza, H., Abdullah, M. A., Khaliq, S., Sajjad, M. H., Elahi, M., Khan, M. R., & Siddique, A. (2024). Role of CRISPR in crop improvement: A review. *Asian Journal of Research in Agriculture and Forestry, 10*(2), 12–19.
43. Chehelgerdi, M., Chehelgerdi, M., Khorramian-Ghahfarokhi, M., Shafieizadeh, M., Mahmoudi, E., Eskandari, F., Rashidi, M., Arshi, A., & Mokhtari-Farsani, A. (2024). Comprehensive review of CRISPR-based gene editing: Mechanisms, challenges, and applications in cancer therapy. *Molecular Cancer, 23*(1), 9.
44. Callaway, E. (2024). 'ChatGPT for CRISPR' creates new gene-editing tools. *Nature, 629*(8011), 272.
45. López Martínez, A., Gil Pérez, M., & Ruiz-Martínez, A. (2023). A comprehensive review of the state-of-the-art on security and privacy issues in healthcare. *ACM Computing Surveys, 55*(12), 1–38.
46. Raab, R., Küderle, A., Zakreuskaya, A., Stern, A. D., Klucken, J., Kaissis, G., Rueckert, D., Boll, S., Eils, R., Wagener, H., & Eskofier, B. M. (2023). Federated electronic health records for the European Health Data Space. *The Lancet Digital Health, 5*(11), e840–e847.
47. Behrouzian Fard, G., Ahmadi, M. H., Gholamin, M., Amirfakhrian, R., Saberi Teimourian, E., Karimi, M. A., & Hosseini Bafghi, M. (2024). CRISPR-Cas9 technology: As an efficient genome modification tool in the cancer diagnosis and treatment. *Biotechnology and Bioengineering, 121*(2), 472–488.
48. Chandrababu, A., & Puthumana, J. (2024). CRISPR-edited, cell-based future-proof meat and seafood to enhance global food security and nutrition. *Cytotechnology*, 1–34.
49. Chanchal, D. K., Chaudhary, J. S., Kumar, P., Agnihotri, N., & Porwal, P. (2024). CRISPR-based therapies: Revolutionizing drug development and precision medicine. *Current Gene Therapy, 24*(3), 193–207.
50. Přibylová, A., & Fischer, L. (2024). How to use CRISPR/Cas9 in plants—From target site selection to DNA repair. *Journal of Experimental Botany*. erae147.

51. Adekugbe, A. P., & Ibeh, C. V. (2024). Advancing healthcare data solutions: Comparative analysis of business and research models in the US. *International Medical Science Research Journal, 4*(4), 373–390.
52. Callahan, A., Ashley, E., Datta, S., Desai, P., Ferris, T. A., Fries, J. A., Halaas, M., Langlotz, C. P., Mackey, S., Posada, J. D., & Pfeffer, M. A. (2023). The Stanford Medicine data science ecosystem for clinical and translational research. *JAMIA Open, 6*(3), ooad054.
53. Samaras, A., Bekiaridou, A., Papazoglou, A. S., Moysidis, D. V., Tsoumakas, G., Bamidis, P., Tsigkas, G., Lazaros, G., Kassimis, G., Fragakis, N., & Vassilikos, V. (2023). Artificial intelligence-based mining of electronic health record data to accelerate the digital transformation of the national cardiovascular ecosystem: Design protocol of the Cardio Mining study. *BMJ Open, 13*(4), e068698.
54. Silvestri, S., Islam, S., Papastergiou, S., Tzagkarakis, C., & Ciampi, M. (2023). A machine learning approach for the NLP-based analysis of cyber threats and vulnerabilities of the healthcare ecosystem. *Sensors, 23*(2), 651.
55. Del Giovane, S., Bagheri, N., Di Pede, A. C., Chamorro, A., Ranallo, S., Migliorelli, D., Burr, L., Paoletti, S., Altug, H., & Porchetta, A. (2024). Challenges and perspectives of CRISPR-based technology for diagnostic applications. *TrAC Trends in Analytical Chemistry*, 117594.
56. Wu, H., Sun, Y., Wang, Y., Luo, L., & Song, Y. (2024). Advances in miniature CRISPR-Cas proteins and their applications in gene editing. *Archives of Microbiology, 206*(5), 231.
57. Eghbalsaied, S., Lawler, C., Petersen, B., Hajiyev, R. A., Bischoff, S. R., & Frankenberg, S. (2024). CRISPR/Cas9-mediated base editors and their prospects for mitochondrial genome engineering. *Gene Therapy, 31*(5), 209–223.
58. Thapar, N., Eid, M. A., Raj, N., Kantas, T., Billing, H. S., & Sadhu, D. (2024). Application of CRISPR/Cas9 in the management of Alzheimer's disease and Parkinson's disease: A review. *Annals of Medicine and Surgery, 86*(1), 329–335.
59. Clayton, E. W., Embí, P. J., & Malin, B. A. (2023). Dobbs and the future of health data privacy for patients and healthcare organizations. *Journal of the American Medical Informatics Association, 30*(1), 155–160.
60. Pillay, R. (2023). Digital health trends. In *Digital Health Entrepreneurship* (pp. 123–129). Cham: Springer International Publishing.
61. Guduri, M., Chakraborty, C., & Margala, M. (2023). Blockchain-based federated learning technique for privacy preservation and security of smart electronic health records. *IEEE Transactions on Consumer Electronics.*
62. Aloini, D., Benevento, E., Stefanini, A., & Zerbino, P. (2023). Transforming healthcare ecosystems through blockchain: Opportunities and capabilities for business process innovation. *Technovation, 119*, 102557.
63. Nowrozy, R., Ahmed, K., Wang, H., & McIntosh, T. (2023). Towards a universal privacy model for electronic health record systems: An ontology and machine learning approach. *Informatics, 10*(3), 60.
64. Thummisetti, B. S., & Atluri, H. (2024). Advancing healthcare informatics for empowering privacy and security through federated learning paradigms. *International Journal of Sustainable Development in Computing Science, 1*(1), 1–6.

65. Wang, W., Li, X., Qiu, X., Zhang, X., Brusic, V., & Zhao, J. (2023). A privacy preserving framework for federated learning in smart healthcare systems. *Information Processing & Management, 60*(1), 103167.
66. Ibeh, C. V., Elufioye, O. A., Olorunsogo, T., Asuzu, O. F., Nduubuisi, N. L., & Daraojimba, A. I. (2024). Data analytics in healthcare: A review of patient-centric approaches and healthcare delivery. *World Journal of Advanced Research and Reviews, 21*(2), 1750–1760.
67. Boopathi, S. (2023). Securing healthcare systems integrated with IoT: Fundamentals, applications, and future trends. In *Dynamics of Swarm Intelligence Health Analysis for The Next Generation* (pp. 186–209). IGI Global.
68. Corpuz, J. C. G. (2023). Metaverse phenomenon and its impact on public health. *Journal of Public Health, 45*(4), e838–e839.
69. Buddha, G. P., & Pulimamidi, R. (2023). The future of healthcare: Artificial intelligence's role in smart hospitals and wearable health devices. *Tuijin Jishu/ Journal of Propulsion Technology, 44*(5), 2498–2504.
70. Medina-de la Cruz, R. O., Gonzalez-Avila, L. U., Saldaña-Padilla, A., Mora-Piña, C. J., Hernández-Cortez, C., & Castro-Escarpulli, G. (2024). CRISPR-based techniques and their application in plants. In *CRISPRized Horticulture Crops* (pp. 91–104). Academic Press.
71. Hajifathali, A., Lasemi, M. V., Mehravar, M., Moshari, M. R., Alizadeh, A. M., & Roshandel, E. (2024). Novelty in improvement of CAR T cell-based immunotherapy with the aid of CRISPR system. *Hematology, Transfusion and Cell Therapy, 46*(1), 58–66.
72. Gratz, S., O'Connor-Giles, K. M., & Wildonger, J. (2024). Generating CRISPR alleles in *Drosophila. Cold Spring Harbor Protocols, 2024*(5), pdb-rot108256.
73. Ansori, A. N., Antonius, Y., Susilo, R. J., Hayaza, S., Kharisma, V. D., Parikesit, A. A., Zainul, R., Jakhmola, V., Saklani, T., Rebezov, M., & Ullah, M. E. (2023). Application of CRISPR-Cas9 genome editing technology in various fields: A review. *Narra Journal, 3*(2), e184–.
74. Stefanoudakis, D., Kathuria-Prakash, N., Sun, A. W., Abel, M., Drolen, C. E., Ashbaugh, C., Zhang, S., Hui, G., Tabatabaei, Y. A., Zektser, Y., & Lopez, L. P. (2023). The potential revolution of cancer treatment with CRISPR technology. *Cancers, 15*(6), 1813.
75. Sauvagère, S., & Siatka, C. (2023). CRISPR-Cas: 'The multipurpose molecular tool' for gene therapy and diagnosis. *Genes, 14*(8), 1542.
76. Bhatia, S., & Yadav, S. K. (2023). CRISPR-Cas for genome editing: Classification, mechanism, designing and applications. *International Journal of Biological Macromolecules, 238*, 124054.
77. Dhasarathan, C., Shanmugam, M., Kumar, M., Tripathi, D., Khapre, S., & Shankar, A. (2024). A nomadic multi-agent-based privacy metrics for e-health care: A deep learning approach. *Multimedia Tools and Applications, 83*(3), 7249–7272.
78. Rasel, M., Bommu, R., Shovon, R. B., & Islam, M. A. (2023). Ensuring data security in interoperable EHR systems: Exploring blockchain solutions for healthcare integration. *International Journal of Advanced Engineering Technologies and Innovations, 1*(01), 212–232.

79. Khang, A., Abdullayev, V., Hrybiuk, O., & Shukla, A. K. (2024). *Computer Vision and AI-Integrated IoT Technologies in the Medical Ecosystem*. CRC Press.
80. Ali, A., Zafar, M. M., Farooq, Z., Ahmed, S. R., Ijaz, A., Anwar, Z., Abbas, H., Tariq, M. S., Tariq, H., Mustafa, M., & Bajwa, M. H. (2023). Breakthrough in CRISPR/Cas system: Current and future directions and challenges. *Biotechnology Journal, 18*(8), 2200642.
81. Chavez, M., Chen, X., Finn, P. B., & Qi, L. S. (2023). Advances in CRISPR therapeutics. *Nature Reviews Nephrology, 19*(1), 9–22.
82. Kumaran, A., Jude Serpes, N., Gupta, T., James, A., Sharma, A., Kumar, D., Nagraik, R., Kumar, V., & Pandey, S. (2023). Advancements in CRISPR-based biosensing for next-gen point of care diagnostic application. *Biosensors, 13*(2), 202.
83. Wei, J., & Li, Y. (2023). CRISPR-based gene editing technology and its application in microbial engineering. *Engineering Microbiology, 3*(4), 100101.
84. Gunitseva, N., Evteeva, M., Borisova, A., Patrushev, M., & Subach, F. (2023). RNA-dependent RNA targeting by CRISPR-Cas systems: Characterizations and applications. *International Journal of Molecular Sciences, 24*(8), 6894.
85. Adeghe, E. P., Okolo, C. A., & Ojeyinka, O. T. (2024). The role of big data in healthcare: A review of implications for patient outcomes and treatment personalization. *World Journal of Biology Pharmacy and Health Sciences, 17*(3), 198–204.
86. Singh, B., & Kaunert, C. (2024). Adventure in high altitude of mountainous topographies and health impacts: Lensing tourism sustainability via reducing ecological and sociocultural footprint and health emergency and medical assistance management. In G. Meraj, S. Hashimoto, & P. Kumar (Eds.), *Navigating Natural Hazards in Mountainous Topographies. Disaster Risk Reduction* (p. 15). Cham: Springer. https://doi.org/10.1007/978-3-031-65862-4_15 (accessed on 5th January 2026).
87. Sindelar, R. D. (2024). Genomics, other "OMIC" technologies, precision medicine, and additional biotechnology-related techniques. In *Pharmaceutical Biotechnology: Fundamentals and Applications* (pp. 209–254). Cham: Springer International Publishing.
88. Singh, B., Singh, A., Kaunert, C., Arora, M. K., Lal, S., & Ravesangar, K. (2025). Unleashing the ethical and legal implications of e-business revolution: Consumer privacy and security concerns in phase of digital disruption. In H. Taherdoost, G. Drazenovic, M. Madanchian, I. Khan, & O. Arshi (Eds.), *Business Transformation in the Era of Digital Disruption* (pp. 157–180). IGI Global. https://doi.org/10.4018/979-8-3693-7056-8.ch006 (accessed on 5th January 2026).
89. Udegbe, F. C., Ebulue, O. R., Ebulue, C. C., & Ekesiobi, C. S. (2024). Precision medicine and genomics: A comprehensive review of IT-enabled approaches. *International Medical Science Research Journal, 4*(4), 509–520.
90. Hashim, J. (2024). Genomic medicine from bench to bedside in clinical practice. *International Journal of Advanced Engineering Technologies and Innovations, 1*(4).

CHAPTER 3

Impact of Genomics on Drug Discovery and Clinical Medicine in Public Healthcare: Fostering Disease Prevention and Enhancing Human Well-Being

ASHIMA JAIN[1*] and BHUPINDER SINGH[2,3,*]

[1]School of Law, Manipal University Jaipur, Jaipur, Rajasthan, India

[2]Sharda School of Law, Sharda University, Greater Noida, Uttar Pradesh, India

[3]University of South Wales, UK; Santo Tomas University Tunja, Colombia

*Corresponding authors

ABSTRACT

With the advent of genomics, drug discovery and clinical medicine have been transformed in public healthcare in the form of precision medicine, early detection of diseases, and personalized therapeutic strategies. By combining genomic data with AI and big data analytics, we can identify potential drug targets more quickly, cutting down the time and cost of drug development. This information is also available in clinical practice with genomic profiling, preventing disease by identifying DNA that could lead to risk, protecting patients through early intervention. These therapeutic approaches have undergone massive developments in recent years, particularly with the emergence of pharmacogenomics that enables the personalization of drug therapies to reduce possible side effects and increase therapeutic effectiveness. Genomics are applied to infectious disease,

Genetic Editing for Healthcare: Advances in Disease Management.
Bhupinder Singh, Rishabha Malviya, Sahil Lal, and Sathvik Belagodu Sridhar (Eds.)

DOI: 10.1201/9781779643827-3

cancer management, and rare genetic disorders, resulting in tremendous improvements in the clinical care of these patients. However, ethical issues, privacy with regard to genetic information, and equitable access to genomic medicine are important challenges which require worldwide policy frameworks. The future of genomics goes through interdisciplinary collaboration with health care, strengthening regulatory frameworks, and technological evolution tailored for bioinformatics. Genomics is transforming modern medicine, making health care more predictive, preventive, and personalized by promoting disease prevention and improving human health.

3.1 INTRODUCTION

Genomics creates a revolution in drug discovery, clinical medicine, and public healthcare by understanding large variations of genetics that drive the disease, providing precision medicine, and developing new therapeutic measures. As the science evolves, genomics-enabled strategies have expeditiously alleviated the drug development process, while fundamentally shaping disease prevention modalities to enhance human life. It represents a transformation from traditional, generic interventions to specific, patient-centric diagnostics and treatments. Genomics is helping to promote proactive healthcare measures that target disease burdens and improve public health outcomes by identifying genetic predispositions to diseases and tailoring therapies accordingly.

With the introduction of genomics to the world of science, molecular-level information about the mechanisms of disease has been made available, and thus, we have entered a new age of drug discovery. Genomic sequencing technologies that enable the identification and validation of targets have revolutionized the way traditional drug discovery relied on cumbersome trial-and-error processes. The beginning of the genomic era, with the Human Genome Project (HGP) and subsequent discoveries based on genomic data, has made it possible to identify genes and genetic defects responsible for many diseases, making drug discovery easier. Pharmacogenomics, the study of how genes impact the way one responds to drugs, is a part of genomics that has resulted in personalized medicine. Reviewing genetic variations enable their identification in a way that doctors could know how patients metabolize drugs, thereby optimizing dosage regimens and minimizing adverse drug reactions. As a result, drug efficacy is increased, and the chance of negative side effects is diminished, leading to improved treatment success. For example, cancer treatment with novel therapies like imatinib and trastuzumab has been made possible by genomic profiling of tumors. Genomics has truly revolutionized oncology with precision medicine strategies that have led to the development of therapies with powerful efficacy targeted to patients' genomic profiles.

3.1.1 BACKGROUND OF GENOMICS IN MEDICINE AND DRUG DISCOVERY

Genomics has enabled the discovery of new druggable targets and has accelerated the design of next-generation therapeutics. Genome-wide association studies (GWAS) and next-generation sequencing (NGS) have facilitated the identification of genetic variants from complex disorders (e.g., diabetes, cardiovascular diseases, neurodegenerative diseases). Developing disease-specific pharmacological agents with a targeted mode of action is now feasible by utilizing this information to create drugs that act against specific genes or pathways implicated in the course of the disease. The successful example is PCSK9 inhibitors, a class of cholesterol-lowering drugs that was discovered after the identification of genetic mutations in the PCSK9 gene in the lower blood cholesterol level. "Genomics-driven drug discoveries like this are a prime example of how genetic research is transforming pharmaceuticals and public health interventions."

3.1.2 GENOMICS IN CLINICAL MEDICINE: PATHWAYS TO PRECISION MEDICINE

Genomics has made it possible to provide personalized and precise healthcare in clinical medicine, leading to major advances in disease diagnosis, prognosis, and therapeutics. This allele provides insight into the genetic make-up of an individual, which allows for the prediction of susceptibility for developing disease before symptoms appear and allows for preventative measures to be used. Such a proactive approach lessens the burden of chronic diseases, improves patient outcomes, and reduces healthcare costs. Genomic testing informs a chief diagnostic and contributes significantly to this preventive medicine domain, considering these diseases are a major factor among leading causes of death in early life. For example, genetic screening for pathogenic *BRCA1* and *BRCA2* mutations can be utilized to identify high-risk individuals at risk for developing breast and ovarian cancer, who can take preventive measures in the form of additional surveillance, lifestyle changes, or prophylactic surgeries. Genomic testing for cystic fibrosis, sickle cell anemia, and Huntington's disease is similarly critical for understanding disease trajectory and management options.

Genomics has transformed infectious disease management through better pathogen detection and surveillance. Application of whole-genome sequencing (WGS) of infectious agents allows for timely detection of antimicrobial resistance genes, leading to targeted treatment strategies. Genomic sequencing has been vital during the COVID-19 pandemic in terms of tracking the evolution of the virus, providing information that guided vaccine development and helping to shape public health recommendations. The use of genomics in epidemiology has improved preparedness and response in outbreaks, therefore protecting global health.

3.1.3 IMPORTANCE OF GENOMICS IN PUBLIC HEALTHCARE

Gene therapy, the modification or replacement of defective genes to treat genetic disorders, is another revolutionary application of genomics in clinical medicine. Applications of gene-editing technologies like CRISPR-Cas9 offer the promise of editing disease-causing mutations in DNA to treat previously unmanageable diseases. Clinical trials show promise in treating genetic conditions ranging from sickle cell disease to muscular dystrophy to some forms of inherited blindness. With the ongoing advances in gene therapies, the expectation is that they will transform the landscape of medicine by offering single, long-lasting cures for genetic conditions. Genomics is transforming disease prevention and population health management based on the application in public healthcare. With the help of genomic medicine, healthcare systems can detect a potential outbreak of disease and take necessary precautions to prevent it. By emphasizing this preventative aspect, individuals will be more invested in treating ailments before they arise, leading to better health and greater longevity.

Public health initiatives based on genomics have been especially productive in newborn screening programs. Numerous nations now conduct genetic screening at birth for hereditary and metabolic disorders to ensure early diagnosis and treatment. For example, conditions like phenylketonuria (PKU), congenital hypothyroidism, and sickle cell disease can be identified and addressed early on, preventing serious complications and enhancing quality of life. The success of these newborn screening programs demonstrates the potential of genomics. It has become an essential tool in studying the genetic component for all major non-communicable diseases (NCDs) that pose the largest threat to humanity in terms of morbidity and mortality worldwide. Facilitated by recent significant advances in DNA sequencing, during the past two decades it has been discovered that many diseases, such as cardiovascular diseases, diabetes, and some cancers, have considerable genetic components that affect susceptibility to and progression of the disease. Governments can design personalized prevention strategies, promote healthy lifestyles, and implement targeted interventions to improve public health policies by employing genomic data [1].

3.2 GENOMICS IN DRUG DISCOVERY: UNDERSTANDING GENETIC VARIABILITY IN DRUG RESPONSE

Genomics has enabled us in drug discovery to understand how the variability of a human genome can affect their response to a medication, with the goal to achieve precision medicine. This has revolutionized the creation of bespoke treatments, in an area of study known as pharmacogenomics, the interaction between an individual's genetic predisposition and the effectiveness of a drug. Conventional drug discovery used trial-and-error approaches, resulting in wide variability of patient response and adverse drug reactions. However, newly developed genomic sequencing technologies have enabled researchers to identify genetic variants that modulate drug-relevant pathways such as metabolism, target engagement, and therapeutic efficacy. Pharmaceutical companies can design drugs by incorporating genomic data, leading to more effective and safer drugs, potentially lowering the risks of adverse drug reactions and inefficacy of treatment.

Genomics has a big impact on drug discovery, in part because of a phenomenon called genetic polymorphisms, or slight variations in DNA sequences that alter the ways in which individual humans process drugs. The class of molecules that show great genetic variation are drug-metabolizing enzymes, particularly those encoded by the cytochrome P450 (CYP) family of genes. Polymorphisms in the *CYP2D6* gene, for instance, affect the metabolism of over 25% of commonly prescribed medicines, including antidepressants, opioids, and beta-blockers. *CYP2D6* mutations result in poor metabolism, which can lead to drug toxicity, whereas ultra-rapid metabolizers may not attain therapeutic drug levels. Such genetic variability can be used to tailor drug doses for an individual, thus enhancing treatment efficacy and decreasing toxic side effects.

Genomics in drug discovery also has another important role in the identification of genetic markers related to drug targets. GWAS and NGS allow researchers to identify pathologic genes and biological pathways associated with diseases. An understanding of these diseases leads to targeted treatments which modulate disease-specific genes or proteins. This is demonstrably the case, as evidenced by the identification of frequent mutations in the EGFR gene in non-small cell lung cancer, resulting in the creation of tyrosine kinase inhibitors (TKIs) such as erlotinib and gefitinib that specifically inhibit EGFR signaling. Just as the presence of genetic mutations such as *BRCA1* and *BRCA2* in breast and ovarian cancer patients has helped target the use of drugs such as the PARP inhibitor olaparib to those most likely to benefit, additional genetic and molecular markers will likewise improve the likelihood that targeted therapies will be effective and that patients will survive.

3.2.1 GENOME-BASED DRUG TARGET IDENTIFICATION

Genomics is also key to drug repurposing, finding new indications for old drugs when the evidence matches that of the new drug. Researchers could identify if existing drugs designed for one condition might work in cases of another by analyzing large-scale genomic datasets. For instance, genetic studies of neurodegenerative diseases have identified common molecular pathways with some cancer types, which has opened new approaches for repositioning cancer drugs for Alzheimer's or Parkinson's disease. This not only speeds up drug discovery but also saves researchers time and money, ultimately getting treatment to patients faster.

Combining artificial intelligence (AI) and machine learning with genomics contributes to an even better drug discovery process, allowing predictive modeling of drug responsiveness according to genetic profiles. AI-enabled genomic analysis can lead to new drug targets, better clinical trial designs, and better patient stratification through genetic susceptibility. Genomics, the study of the human genome to find molecular patterns that might serve as markers for disease, is constantly evolving, and its increasing involvement in drug discovery will lead to safer, more effective therapeutic solutions that are tailored specifically to an individual, which is a hallmark of precision medicine.

3.2.2 ROLE OF ARTIFICIAL INTELLIGENCE (AI) AND BIG DATA IN GENOMIC DRUG DEVELOPMENT

With the assistance of big data, genomic drug development has been simplified, with numerous sequenced genomic data analyzed in a much shorter timeframe; AI and Big Data in Genomic Drug Development. Traditional drug development can take more than a decade and cost over a billion dollars to bring a drug to market. But AI-guided computational models and Big Data analytics have revolutionized the field by discovering genetic markers, predicting drug responses, and optimizing the designs of clinical trials. Such artificial intelligence training enables researchers to analyze millions of genomic sequences in an effort to identify patterns and correlations that would have been impossible to detect using traditional methodologies. Incorporating AI and Big Data into genetics to convert pharmaceutical companies to more focused caring medicines, cheaper drug development, and better final results.

There are multiple ways AI can contribute to drug development based on genomics. One of them is discovering new drug targets. A genome is sequenced to find out the disease genes, which is done by machine learning algorithms. AI-enabled methods, such as deep learning algorithms, analyze genomic sequences for tracing mutations, structural variations, and epigenetic alterations associated with the diseases. AI has also facilitated the identification of specific mutations in cancer genomes; for example, mutations in classical oncogenes such as *TP53* and *KRAS* have enabled the design of precision therapies directed toward these genetic alterations.

Big Data plays a vital role in another key aspect of pharmacogenomics, which is understanding genetic variability in drug response. Using genomic, transcriptomic, and proteomic data from different populations, AI models can predict how different genetic backgrounds impact drug metabolism and efficacy. This has also helped optimize personalized therapies for diseases such as cancer, cardiovascular diseases, and neurological disorders. AI improves clinical trial efficiency by automating the patient selection process and monitoring treatment responses in real time. These AI-powered platforms can review the electronic health records (EHRs), genomic profiles, and lifestyle data of millions to identify the right fit for clinical trials, making sure the relevant genetically aligned populations participate in trials. AI-harnessed predictive analytics can also capture and forecast drug toxicity and efficacy, allowing researchers to optimize drug formulations before they even get to human trials. As AI and Big Data progress, their integration in genomic drug development will enable faster, cheaper, and ultra-personalized treatments, fundamentally revolutionizing public healthcare and promoting precision medicine and enhanced disease prevention.

3.3 CLINICAL APPLICATIONS OF GENOMICS: GENOMIC BIOMARKERS IN DISEASE DIAGNOSIS

Since the identification of genomic biomarkers, disease diagnosis has improved significantly due to early detection, prognosis, and personalized medication approaches. The evolving field of genomics has enabled scientists to discover particular genetic variants, mutations, and epigenetic changes linked to different diseases, which can facilitate more accurate and early diagnoses. Typical diagnostic methods depend on clinical symptoms and biochemical indicators that can lack specificity and potentially slow the delivery of treatment. Genomic biomarkers, on the other hand, offer molecular-level information on disease biology, enabling early diagnosis and better-informed treatment choices. This has been especially successful for oncology, neurology, cardiovascular, and infectious diseases, where genomic biomarkers help classify patients into risk groups, predict disease evolution, and optimize therapies.

Genomic biomarkers are transforming oncology. These testing platforms identify the genomic mutations that fuel tumor growth, thus allowing for personalized cancer treatment. Mutations in the *BRCA1* and *BRCA2* genes, for example, markedly raise the risk for breast and ovarian cancer. Detecting these mutations with genetic testing enables high-risk people to take preventive steps, including more aggressive screening or prophylactic surgeries. They have been used as biomarkers to guide therapeutic agency in lung cancer, such as mutations of the *EGFR*, *ALK*, and *KRAS*. For example, those with EGFR mutations are effectively treated with TKIs, including erlotinib and gefitinib, that selectively inhibit aberrant EGFR signaling cascades, thus increasing therapeutic efficacy and survival rates.

Genomic biomarkers also play a key role in making common neurological disease diagnoses, such as neurodegenerative diseases including Alzheimer's and Parkinson's disease. APOE genetic variations, particularly the APOE ε4 allele, have a strong individual correlation with the risk of Alzheimer's disease. Identifying such risk factors at an early stage allows both patients and clinicians to undertake lifestyle changes and interventions that might postpone the onset of disease. In parallel, mutations to the *LRRK2* and *SNCA* genes have been characterized as biomarkers of Parkinson's disease that can also facilitate differential diagnosis and the potential for gene-targeted therapies. In cardiovascular diseases, genomic biomarkers are used to assess a person's risk of conditions such as coronary artery disease, arrhythmias, hypertension, and more. Mutations in the *PCSK9* gene, for example, affect a response in cholesterol metabolism, which led to the development of new types of cholesterol-lowering agents, including *PCSK9* inhibitors. Likewise, mutations in the *KCNQ1* and *SCN5A* genes are associated with congenital long QT syndrome, a condition that predisposes individuals to potentially fatal arrhythmias. Detecting these variants can lead to early interventions, whether through lifestyle changes, medication, or implantable defibrillators, to avert sudden cardiac events.

Genomic biomarkers are crucial for the diagnosis and management of infectious diseases. Genomic sequencing for pathogens is now enabling rapid identification of viral and bacterial strains, which can guide antimicrobial therapy and control of outbreaks. SARS-CoV-2 genomic analyses during the coronavirus disease 2019 (COVID-19) pandemic enabled the monitoring of viral mutations to understand immune escape, eventually predict vaccine efficacy, and develop strategies to inhibit viral-mediated pathologies. The host genetic factors, including polymorphisms in the *HLA* and *ACE2* genes, are implicated in varied susceptibility to infectious agents and may inform personalized treatment strategies. This will also facilitate the collaborative exchange between genomics and other biological fields, highlighting the potential for more precise, proactive, and personalized approaches to disease diagnosis, facilitated by genomic biomarkers as genomic technologies continue to evolve.

3.3.1 PERSONALIZED MEDICINE AND PRECISION HEALTHCARE

Personalized medicine and precision healthcare are at the frontier of modern medicine, adjusting therapies to an individual's genetic, environmental, and lifestyle circumstances. Personalized medicine uses genomic data, biomarkers, and advanced analytics to create targeted therapies tailored to individual patients, enhancing treatment response and minimizing adverse drug events. Advanced techniques such as AI and big data analysis have increased the strength of precision healthcare such that expansive datasets can be analyzed to estimate disease risk, tailor drug selection, and create individualized treatment plans [2].

The most prominent examples of personalized medicine are in oncology, in which genetic profiling of tumors enables clinicians to tailor targeted therapies to patients' specific mutations. As an illustration, patients with HER2-positive breast cancer respond well to treatment with trastuzumab (Herceptin), and TKIs, such as erlotinib or gefitinib, are used in patients with EGFR mutations in lung cancer. These precision treatments improve survival rates and decrease needless side effects. In pharmacogenomics, genetic testing allows doctors to determine the best dosages and types of drugs for a patient to take. For instance, gene variations in the CYP2C19 enzyme affect patients' metabolism of clopidogrel, a widely used blood thinner, allowing for effective clot prevention without overdose risks. For instance, genotyping for transformation of the TPMT enzyme can guide thiopurine drug dosing for leukemia and autoimmunity. In addition to genetics, precision healthcare accounts for lifestyle and environmental factors, leveraging wearable devices, electronic health records, and AI-enhanced predictive analytics to track real-time health metrics. This allows for early disease identification, prompt preventive transitions, and personalized wellness methods. As personalized medicine progresses, the combination of genomics, AI, and digital health technologies will continue to improve patient outcomes and lower healthcare costs, leading toward a more proactive, data-driven healthcare system.

3.3.2 ROLE OF GENOMICS IN CANCER THERAPY

This makes genomic data fundamentally different than, for example, a privacy binary, For example, the data available for the common remote. Genomic-driven treatments, apart from traditional chemotherapy, differ in the latter's lack of distinction between cancerous and healthy cells, while genomic-driven approaches target personalized interventions based on a patient's genetic constitution. NGS and comprehensive bioinformatic tools have enabled clinicians to characterize tumor-specific mutations and optimize targeted therapies, immunotherapies, and personalized treatment plans that enhance patient outcomes and reduce unwanted side effects.

Genomic-targeted therapy is the most important step of genomic cancer-targeted therapy, which is based on the fact that drugs are developed to block certain molecular pathways involved in the progression of cancer.

HER2-positive breast cancer patients receive trastuzumab (Herceptin), which targets the overexpressed HER2 protein in their own tumors. Likewise, TKIs like erlotinib or osimertinib are prescribed to patients with non-small cell lung cancer (NSCLC) containing mutations to the EGFR gene, inhibiting aberrant EGFR signaling and lessening tumor growth. For instance, BRAF inhibition with vemurafenib has also been a major success in melanoma, as shown by genomic sequencing that identified BRAF V600E mutations in some patients, leading to significantly improved outcomes [3].

Aside from targeted therapies, genomics also has a major role in immunotherapy in regard to predicting patient response to immune checkpoint inhibitors. PD-L1 expression levels and tumor mutational burden (TMB) are profiled biomarkers to identify cohorts of patients that would respond to selected immune checkpoint inhibitors such as pembrolizumab (Keytruda) and nivolumab (Opdivo). CAR-T cell therapy, a revolutionary form of immunotherapy where T-cells are genetically engineered from a patient's body to target and kill cancer cells, has demonstrated significant efficacy in blood cancers such as lymphoma and leukemia.

Genomics also plays a role in liquid biopsies, a technique that allows for the identification of circulating tumor DNA (ctDNA) from a blood sample. This approach enables the tracking of tumor evolution in real time, early detection of mutations conferring drug resistance, and adaptation of the therapy in a timely manner. For instance, EGFR T790M mutations in lung cancer patients indicate resistance to first-line TKIs, which are then replaced by second- or third-generation inhibitors such as osimertinib. Genomic profiling may also predict chemotherapy sensitivity and resistance, which ultimately aids in optimizing treatment regimens tailored to each patient. As a case in point, *BRCA1/2* mutations predict sensitivity to PARP inhibitors in both ovarian and breast cancer, while TP53 mutations may predict poor outcome through resistance to standard of care chemotherapies. With continued advancements in genomic technologies, gene therapy will be incorporated into cancer therapies which would continue to optimize personalized treatment approaches, reduce healthcare costs, improve long-term survival rates and transform the future of oncology.

3.4 DISEASE PREVENTION THROUGH GENOMICS

A key component of genomics in disease prevention is being able to understand one's genetic risk factors, helping us achieve early intervention, which can potentially help in preventing future diseases. Scientists can identify inherited mutations and susceptibilities to a wide range of diseases by studying a person's genome, enabling them to make early medical decisions that can help prevent and attenuate disease. Most existing preventative health systems that are based on biochemistry or health records work reactively when sickness arises, but genomics lets us work much more proactively since we can use a person's genomic data to reveal their potential for developing various health conditions.

Genomics may arguably have the biggest impact on disease prevention in the context of hereditary cancer syndromes [4]. Testing for mutations in genes like *BRCA1* and *BRCA2* can identify people at high risk for breast and ovarian cancer. One such preventive measure would be increased surveillance, chemoprevention, or prophylactic surgeries that can reduce cancer risk significantly. In a similar manner, identifying Lynch syndrome, a hereditary condition caused by mutations in DNA mismatch repair genes, allows for early colorectal cancer screening and preventive measures for individuals and their families.

Genomics is playing an important role in the prevention of cardiovascular disease as well. Mutations in genes (such as *PCSK9*, *APOE*, and *KCNQ1*) have been associated with increased risks for hypercholesterolemia, hypertension, and arrhythmias. The recognition of these variants enables early lifestyle changes, targeted drug interventions, and screening for early cardiovascular disease. In the field of infectious diseases, genomics helps in prediction of susceptibility to infections and also in optimizing vaccine responses. And differences in the HLA and ACE2 genes shape individual responses to viral infections, including COVID-19, informing personalized prevention strategies. Genomics helps enhance nutrigenomics, the study of how genes impact dietary outcomes. Targeted Nutrition (TN) can be designed based on individual genome prior to onset of metabolic disorders for example, diabetes, obesity, and so forth. The deployment of genomic technologies into the mainstream of healthcare will drive a preventive model creating lower disease burden and improved public health as the genomic revolution unfolds.

3.5 ETHICAL, LEGAL, AND SOCIAL CONSIDERATIONS IN GENOMIC MEDICINE

Genomic medicine has transformed the healthcare landscape, allowing for personalized treatment strategies, disease prediction, and early interventions.

But, together with its tremendous promise, come major ethical, legal, and social issues that will have to be confronted to facilitate responsible and equitable use. In this era of increasing importance of genomics, ethical issues including genetic privacy, informed consent, data security, discrimination in the field of genomics, as well as equitable access to genomic healthcare have become the key determinants of this future.

Genomic privacy and confidentiality is the prominent ethical issue in genomic medicine. Genomic data hold very sensitive personal information that, if used or released without consent, may expose individuals to discrimination or stigmatization. This could lead to genetic discrimination, where insurance companies, employers, or other third parties gain access to personal genetic data without consent, and then deny individuals coverage or employment based on their genetic susceptibility to disease. While legislation like the Genetic Information Nondiscrimination Act (GINA) protects against such discrimination in the U.S., similar measures are either weak or nonexistent elsewhere in the world [5].

A related ethical challenge is informed consent for both genomic research and clinical care. Since genomic data and its implications are so complex, patients need to be fully aware of possible risks, benefits, and future use of their genes before giving consent for testing or data sharing. But many people will not have the genomic literacy to make adequately informed decisions. In addition, WGS technology and polygenic risk score technologies are rapidly improving, and incidental findings of unrelated but clinically relevant genomic variants may arise that give rise to conflicting, deontological quandaries about whether or how such findings should be disclosed to patients.

Legal issues regarding genomics data protection and ownership are considerable hurdles. Genomic data, on the other hand, is often in a regulatory gray area, especially when in large databases or used for research, while traditional medical data is protected by strict confidentiality laws. Biobanking, the collection and storage of genetic material for studies not yet realized, raises questions about who owns genetic data and who gets to use it. The use of genetic research raises ethics questions about how participant consent is handled, creating issues where participants may unknowingly give away the rights to their research when donating their genetic material.

Equitable access to genomic technologies and better healthcare is a critical social issue affecting genomic medicine. Although precision medicine provides potentially life-changing treatments for these individuals, their price and inaccessibility to underserved populations are further widening already staggering health disparities. High-income countries have better access to genetic testing and personalized treatments, while cost, infrastructure, and regulatory obstacles plague low- and middle-income countries. This global divide indicates ethical dilemmas regarding fairness and justice in healthcare delivery, and conveys the necessity of policies to enable affordable, equitable access to genomic-based innovations.

3.5.1 ETHICAL CHALLENGES IN GENOMIC RESEARCH AND CLINICAL USE

AI and big data in genomic research pose several new ethical and legal issues. AI-based algorithms are applied to analyze extensive genomic datasets, forecast disease risks, and direct precision therapies [6]. But algorithmic bias and data privacy issues can be dangerous. Models trained on non-representative data risk giving inaccurate or biased results for under-represented ethnic or genetic population groups, perpetuating health inequalities. The transparency and accountability of AI-driven genomic decision-making remains an urgent legal and ethical problem.

The ethics of gene editing technologies such as CRISPR-Cas9 is a similarly important debate. Genome editing offers hope of curing inherited diseases during embryonic development but also poses ethical challenges around human genetic enhancement and unintended genetic alterations. Editing human embryos to eradicate the risk of disease or to enhance physical and cognitive characteristics is a prospect that has stirred controversy worldwide, raising fears of eugenics, social divisions, and unforeseen genetic mutations. Many countries have implemented tight restrictions on germline editing, but global agreement on where ethical lines should be drawn remains lacking. Cultural beliefs and religious practices need to be factored when discussing genomic medicine. Genetic testing, gene editing, reproductive choices—each of these come with different beliefs in different communities. For example, some religions may reject genetic changes for reasons outside of ethics; other concerns related to their faith or a community may not accept genetic testing for inherited diseases for cultural reasons. These intrinsic beliefs must be explicitly acknowledged and addressed through culturally appropriate communication and engagement with communities, ensuring ethical practices involving genomic healthcare.

The other industrial challenge to ethical and legal landscapes has emerged, namely the rise of direct-to-consumer (DTC) genetic testing, which is offered by companies. Although these tests can sometimes offer information about ancestry, disease risk, and carrier status for specific diseases, they often operate without the guidance of a medical professional, and genetic results can be widely misinterpreted [7]. The lack of DAC verification can lead to consumers making potentially important medical choices based on incomplete or inaccurate information, which is why genetic counseling and regulatory oversight of the DTC genetic testing industry are so important. Genomic medicine offers great potential for enhancing health care; however, it also raises challenging ethical, legal, and social issues that demand careful consideration. The responsible advancement of genomic technologies relies on genetic privacy protection, informed consent rights, prevention of discrimination, equity of access, and the establishment of legal frameworks. As genomic medicine evolves, interdisciplinary collaboration among scientists, policymakers, ethicists, and legal experts will be crucial to striking the right balance between innovation and ethical responsibility, laying the groundwork for a more equitable and trustworthy genomic healthcare system.

3.5.2 DATA PRIVACY AND SECURITY IN GENOMICS

Their training allows them to understand the nuances of genomic data and its clinical application, facilitating genomics-enabled therapies, precision medicine, and population genomics at scale. Yet, socioeconomic inequalities in access to genomic technologies continue to represent an important problem, worsening actual healthcare inequities [8]. The accessibility of genomic medicine has been uneven in low- and middle-income populations due to factors such as high costs, limited healthcare infrastructure, low genomic literacy, and unequal distribution of research and research resources. These inequalities influence not just individual health, but social behavior, health policy direction, and rivalries in public health development, which serve as a systematic perpetuation of inequality in global healthcare.

Cost and affordability are the major barriers to the implementation of genomic medicine. WGS and precision-targeted therapies are exceedingly expensive, and therefore out of reach for lower-income individuals. In wealthy countries, private insurance or government-sponsored programs may pay for genomic tests, but in much of the world, patients cannot afford to pay out of pocket for genetic and precision medicine diagnostics and treatment. This would leave the poor and socioeconomically disadvantaged groups dependent on regularly utilized but less effective treatment alternatives, while only the rich have the capacity to capitalize on their genomic innovations.

Healthcare infrastructure and resource allocation are critical factors in determining access to genomic medicine. Developed nations have established genomic research labs, high-tech laboratories, and specialist physicians to interpret genomic data and provide personalized healthcare. In the less wealthy countries, access to sequencing machines, trained geneticists, and bioinformatics resources are scant. Many populations in these regions may be deprived of the latest advances of genomic medicine due to insufficient investment in genomic research and healthcare infrastructure [9].

3.5.3 SOCIOECONOMIC DISPARITIES IN ACCESS TO GENOMIC MEDICINE

Genomic literacy and awareness are another big driver of disparity. Here is the human-like style version of it. Due to ignorance and lack of knowledge, there is comparatively lesser participation in the genetic screening program and hence a delayed diagnosis and missed opportunity of preventive healthcare. Such access is also thwarted by cultural and language barriers; some groups may also be skeptical or misinformed about the health implications of genetic testing for religious, ethical, or historical reasons [10]. To address this gap in genomic knowledge, public health efforts should emphasize community-specific education tailored to the target population.

Lack of diversity in genomic research participation is another cause of inequity in access. Genomic studies, historically, have tended to over-represent populations of European ancestry, and as a result, genetic databases that drive prospective precision treatments are not as effective in ethnically diversified peoples. Lack of data on their unique heredity may mean that genomic discoveries do not benefit them, and underrepresented populations risk being left behind in the genomic revolution [11]. This expanding participation in genomic research is critical for equitable progress in precision medicine.

Policy and regulatory environments are also central in shaping equitable access to genomic medicine. In others, there is no clear guidance as to the regulatory status of genomic diagnostics, or public healthcare funding is limited to a few conditions, disadvantaging those at risk but who cannot get insurance coverage in a private healthcare system, or who do not have the supportive background for affordability of options. While countries with organized universal healthcare systems are in better shape to embed genomic medicine into standard practice, those with privatized or otherwise fractured healthcare models often experience extremes of inequality in access. As genomic medicine accounts for an increasingly important contribution to the prevalence and treatment of diseases, the development of global health initiatives must disregard costly gaps in education, infrastructure, and accessibility to progress in genomics and genetic engineering [12]. Yet genomic medicine offers undoubted promise in improving maternal health outcomes, and governments, healthcare systems and private organizations must cooperate to ensure that both research and genomic interventions reach all populations rather than only the privileged few. These efforts include subsidizing genetic testing for low-income patients, expanding genomic research to include diverse populations, increasing availability of genetic counseling services, and strengthening healthcare systems in underserved regions.

Genomic medicine is extremely promising, yet socioeconomic disparities could pose a barrier to accessibility. In addressing these inequities, we need to be multidisciplinary, decrease cost, build up infrastructure, educate the public, include diverse representation in research, and devise strong policy. Equity in genomic healthcare promotes health equity globally, whereby the precision medicine available in genomic healthcare promotes accessibility and affordability, particularly for individuals in economically challenging circumstances.

3.6 GENOMICS AND PUBLIC HEALTH POLICIES: GLOBAL INITIATIVES IN HEALTHCARE

It is through genomics that we have reformed healthcare by enhancing early disease detection, tailoring treatment approaches, and supporting public health efforts. So, as genomic technologies develop, public health policies must evolve to safeguard that such advancements preserve populations in a fair and effective way. Initiatives by governments, international organizations, and health-care systems around the world are based on these capabilities and have been launched or are in development to integrate genomics into public health through prevention of disease, precision medicine, aggregation of data, and ethics. Global genomic healthcare remains in its infancy, and addressing these challenges will be critical for establishing equitable genomic healthcare policies worldwide [13].

Genomics plays a major role in modern medicine, especially in public health policies, especially in disease prevention and early detection programs. Around the world, many governments have introduced initiatives aimed at incorporating genomic screening into their National Health Systems to identify at-risk patients for genetic disorders, cancers, and other hereditary diseases. For example, both the 100,000 Genomes Project in the U.K. and the Precision Medicine Initiative in the U.S. aim to harness the power of the genome to predict, prevent, and treat diseases more effectively. These projects allow healthcare providers to create customized prevention measures, leading to a lower overall burden of chronic diseases. Newborn screening for genetic disorders using genomic data is already being greatly expanded in several nations, enabling early detection and management of these conditions and leading to improved health outcomes.

On the other hand, the inclusion of genomic medicine into national healthcare policies demands effective regulatory frameworks that address data security, ethical inquiries, and fair access. Genomic data privacy is protected under Section 9 of the General Data Protection Regulation (GDPR) of the European Union. Likewise, in the U.S., the GINA protects people from genetic discrimination from employers and insurance companies [14]. Yet, there are few genomic regulations in LMICs, which also raises concerns around genetic privacy, data security, and stigmatization through misuse of genomic data. To ensure that genomic health care policies do not undermine ethical practice and trust people must have in their care, alignment around data protection laws by developing international governance models is key.

3.6.1 POLICY FRAMEWORKS FOR INTEGRATING GENOMICS INTO PUBLIC HEALTH

Data sharing and worldwide collaboration is another crucial component of global genomic efforts. Genomic data is helpful for any discoveries, from identifying disease patterns to developing tailored solutions or improving precision medicine [15]. Data sharing between researchers is possible as there are initiatives like the Global Alliance for Genomics and Health (GA4GH), the Human Genome Project, and so forth, which promotes innovation in healthcare. Nonetheless, regional data must be standardized, and issues relating to data control, patient opt-in, and equitable distribution of benefits from shared discoveries are not yet solved. This sharpens some of the ethical and legal boundaries that need to apply to the sharing of genomic data to protect against misuse, but also to enable scientific progress.

Public health policies encompass not only the integration of genomics into primary healthcare, ensuring access to the majority. Data reported to the World Health Organization (WHO) highlights the importance of incorporating genomic medicine into national health policy Directions to Develop the Genomic Medicine State of the Union: Infectious Diseases (includes prevention, response, and treatment), Cancer, and Rare Diseases. An example of this importance occurred during the COVID-19 global pandemic when genomic sequencing was instrumental in tracking virus mutations, developing vaccines, and informing public health responses [16]. Countries that had already invested in genomic surveillance systems were capable of responding more effectively to emerging variants, demonstrating the need to integrate genomic capabilities into national health infrastructures.

3.6.2 GENOMIC MEDICINE IN HEALTHCARE SYSTEMS

But there are significant socioeconomic and geographic differences in the availability of genomic medicine. LMICs face constraints in access to genetic testing, trained personnel, and health funding compared to high-income countries with an active genomics and precision medicine landscape. To fill this gap, international organizations like the WHO, the Bill and Melinda Gates Foundation, and the trust are investing resources in genomic research and infrastructure in developing areas. Initiatives such as H3Africa (Human Heredity and Health in Africa) are helping to enhance the capacity for genomic research in Africa, thus allowing diverse populations to be included in global genomic data sets and to participate in advances in precision medicine [17].

Genomic medicine continues to raise ethical considerations integral to public health policies. The key issues of inclusion genomics and building public trust in genomics include technologies-centric thinking through which genomics healthcare must not be provided equitably to all people irrespective of their or other social implications, including informed consent and cultural considerations. Indigenous and minority communities have long expressed insecurities regarding genetic exploitation, past misuse of genetic information, and absence from research. Policymakers should certainly strive to confirm that engagement in a community, along with oversight of ethics, receives attention in planning genomic programs. Other interventions include education and awareness efforts to promote understanding of genomic medicine, minimize misinformation, and promote participation in genetic research. Future public health policies should prioritize the provision of genomic literacy, scientific collaboration, robust data security, and equitable accessibility to genomic healthcare [18]. Such regulatory frameworks must balance the dual imperatives of innovation and ethical responsibility and will require joint efforts between governments and international organizations. This can now, finally, become a reality for every single population because we know it has to invest in genomic research, healthcare infrastructure, and workforce training to translate genomic discoveries into tangible health for everyone.

Genomics has revolutionized public health policy on this planet, providing unique opportunities for effective disease prevention, tailor-made treatment, and enhanced patient care. To successfully incorporate genomic medicine into public health, global initiatives must navigate and overcome regulatory voids, socioeconomic differences, and ethical questions to ensure equitable and ethical implementation. Through our work, it can contribute to healthier societies and promote global health sustainable through international collaboration, equitable access, and ethical governance [19].

3.7 CONCLUSION AND FUTURE OF PERSONALIZED GENOMIC HEALTHCARE

While genomics has the potential to revolutionize medicine, many factors are hampering widespread adoption in public healthcare, particularly aspects of ethics, law, and society. To responsibly and ethically implement genomics technology, several concerns like genetic privacy, data security, and equal access to service must be rectified. In addition, inequities in access to genetic testing and precision medicine continue to be an issue, especially in low-resource contexts. Closing these gaps will need the work of governments, healthcare services, as well as policymakers to create fair genomic healthcare models. Besides disease prevention, population health research has been facilitated by genomics studies that have enabled the discovery of genetic variations affecting drug responses in various ethnicities. Pharmacogenomics studies have shown that accounting for genetic diversity can help inform prescribing habits, leading to more effective and safe medication use across populations. Such insight has guided regulatory recommendations and is changing the paradigm of individualized therapy to better incorporate public health measures into clinical practice.

3.8 FUTURE DIRECTIONS

Genomics has now become a cornerstone of drug discovery, clinical medicine, and public healthcare, and its contributions will only become more prominent, accelerated by advances in AI, big data analytics, and precision medicine. The integration of AI with genomic analysis is speeding up the identification of drug targets, making it easier to design the best clinical trials and predicting which patients will benefit from which treatments with breeding accuracy. These innovations have the potential to create a paradigm shift in the state of the art in this clinical field, thereby bringing precision medicine within reach for many more patients. The integration of genomics with other omics technologies, including proteomics, metabolomics, and lipidomics, will help to achieve a more complete understanding of disease mechanisms, leading to the development of more comprehensive therapeutic strategies that can target multiple pathways involved in disease pathogenesis. As genomic research pathways compound, its applications will become even more precise in terms of disease prevention, early diagnosis, and individualized treatment, ultimately benefiting human well-being at a planetary scale. Genomics has revolutionized drug discovery, clinical medicine, and public health care, allowing for more personalized medications, better prevention methods, and improved health outcomes overall. From precision medicine to pathogen surveillance to genetic research, its impact has reshaped healthcare paradigms to create targeted, effective, and patient-centric medicine. Nevertheless, tackling ethical concerns, promoting equitable access, and facilitating interdisciplinary partnerships will be crucial to fully leverage the potential of genomics in healthcare. With the implementation of these advancements, societies can be better positioned towards a future of genomic-driven healthcare that empowers individuals, bolsters public health, and enhances long-term well-being.

KEYWORDS

- **big data analytics**
- **bioinformatics**
- **drug discovery**
- **genetic disorders**
- **genetic variability**
- **genomic biomarkers**
- **genomics**
- **pharmacogenomics**

REFERENCES

1. Marques, L., Costa, B., Pereira, M., Silva, A., Santos, J., Saldanha, L., Silva, I., Magalhães, P., Schmidt, S., & Vale, N. (2024). Advancing precision medicine: A review of innovative *in silico* approaches for drug development, clinical pharmacology and personalized healthcare. *Pharmaceutics, 16*(3), 332.
2. Olorunsogo, T. O., Balogun, O. D., Ayo-Farai, O., Ogundairo, O., Maduka, C. P., Okongwu, C. C., & Onwumere, C. (2024). Bioinformatics and personalized medicine in the US: A comprehensive review: Scrutinizing the advancements in genomics and their potential to revolutionize healthcare delivery. *World Journal of Advanced Research and Reviews, 21*(1), 335–351.
3. Horgan, D., Pesapane, F., Van der Buckle, M., de Maria, R., Dube, F., Singh, J., Ługowska, I., Bayle, A., Hofman, P., Malapelle, U., & Hills, T. (2024). From theory to practice: Implementing next-generation sequencing and public health genomics in healthcare systems. *Critical Reviews in Oncology/Hematology, 201*, 104433.
4. Chafai, N., Bonizzi, L., Botti, S., & Badaoui, B. (2024). Emerging applications of machine learning in genomic medicine and healthcare. *Critical Reviews in Clinical Laboratory Sciences, 61*(2), 140–163.
5. Paraskevas, T., Papapanou, M., Sergentanis, T. N., Kyriopoulos, I., & Athanasakis, K. (2025). Comprehensive genomic profiling: A public health system perspective. *Expert Review of Molecular Diagnostics*. (just-accepted).

6. Carini, C., & Seyhan, A. A. (2024). Tribulations and future opportunities for artificial intelligence in precision medicine. *Journal of Translational Medicine, 22*(1), 411.
7. Liu, H. (2025). Implementations of quantum computing in medicine: Evidence from drug discovery, genomics, and medical imaging. *Highlights in Science, Engineering and Technology, 128*, 238–247.
8. Rasool, S., Ali, M., Shahroz, H. M., Hussain, H. K., & Gill, A. Y. (2024). Innovations in AI-powered healthcare: Transforming cancer treatment with innovative methods. *BULLET: Journal Multidiscipline Ilmu, 3*(1), 118–128.
9. Kokudeva, M., Vichev, M., Naseva, E., Miteva, D. G., & Velikova, T. (2024). Artificial intelligence as a tool in drug discovery and development. *World Journal of Experimental Medicine, 14*(3), 96042.
10. Anyanwu, E. C., Arowoogun, J. O., Odilibe, I. P., Akomolafe, O., Onwumere, C., & Ogugua, J. O. (2024). The role of biotechnology in healthcare: A review of global trends. *World Journal of Advanced Research and Reviews, 21*(1), 2740–2752.
11. Zeb, S., Nizamullah, F. N., Abbasi, N., & Fahad, M. (2024). AI in healthcare: Revolutionizing diagnosis and therapy. *International Journal of Multidisciplinary Sciences and Arts, 3*(3), 118–128.
12. Jeyaraman, N., Jeyaraman, M., Yadav, S., Ramasubramanian, S., & Balaji, S. (2024). Revolutionizing healthcare: The emerging role of quantum computing in enhancing medical technology and treatment. *Cureus, 16*(8).
13. Ishimaru, S., Shimoi, T., Sunami, K., Nakajima, M., Ando, Y., Okita, N., Nakamura, K., Shibata, T., Fujiwara, Y., & Yamamoto, N. (2024). Platform trial for off-label oncology drugs using comprehensive genomic profiling under the universal public healthcare system: The BELIEVE trial. *International Journal of Clinical Oncology, 29*(2), 89–95.
14. Li, J., Wang, S., Rudinac, S., & Osseyran, A. (2024). High-performance computing in healthcare: An automatic literature analysis perspective. *Journal of Big Data, 11*(1), 61.
15. Struelens, M. J., Ludden, C., Werner, G., Sintchenko, V., Jokelainen, P., & Ip, M. (2024). Real-time genomic surveillance for enhanced control of infectious diseases and antimicrobial resistance. *Frontiers in Science, 2*, 1298248.
16. Venkatesh, K. P., Brito, G., & Kamel Boulos, M. N. (2024). Health digital twins in life science and health care innovation. *Annual Review of Pharmacology and Toxicology, 64*(1), 159–170.
17. Yadav, S., Singh, A., Singhal, R., & Yadav, J. P. (2024). Revolutionizing drug discovery: The impact of artificial intelligence on advancements in pharmacology and the pharmaceutical industry. *Intelligent Pharmacy.*
18. Yang, X., Huang, K., Yang, D., Zhao, W., & Zhou, X. (2024). Biomedical big data technologies, applications, and challenges for precision medicine: A review. *Global Challenges, 8*(1), 2300163.
19. Goswami, S., & Sharma, S. (2024). Artificial intelligence, quantum computing and cloud computing enabled personalized medicine in next generation sequencing bioinformatics. In *2024 IEEE International Conference on Interdisciplinary Approaches in Technology and Management for Social Innovation (IATMSI)* (Vol. 2, pp. 1–5). IEEE.

CHAPTER 4

Challenges and Approaches to Human Genome Editing

SAQUIB AHMED[1] and BHUPINDER SINGH[2*]

[1]*Sharda School of Law, Sharda University, Greater Noida, Uttar Pradesh, India*

[2]*Sharda School of Law, Sharda University, Greater Noida, India; University of South Wales, UK; Santo Tomas University Tunja, Colombia*

*Corresponding author

ABSTRACT

Human genome editing has the potential to revolutionize human health. Fields like human health, animal, and plant breeding, environmental applications, and fundamental biomedical research are already using genome editing, and many more are expected to do so in the future. The revolutions can be seen in the field of human genome editing. Human genome editing has the potential to completely transform health care. Genome editing is being used in many different fields, including basic biomedical research, animal, and plant breeding, human health, and environmental applications. This chapter focuses on human genome editing, specifically highlighting recent initiatives and suggested long-term plans for the worldwide regulation of this technology. Traditional government control through laws, rules, and treaties is included in governance, along with more informal supervision and direction from nongovernmental organizations like trade associations, professional associations, civil society organizations, journals, funders, and other outside parties.

Genetic Editing for Healthcare: Advances in Disease Management.
Bhupinder Singh, Rishabha Malviya, Sahil Lal, and Sathvik Belagodu Sridhar (Eds.)

DOI: 10.1201/9781779643827-4

4.1 INTRODUCTION

Genome editing is being used in many different fields, including basic biomedical research, animal, and plant breeding, human health, and environmental applications [1]. In the future, gene editing is expected to be used significantly more in each of these fields. The ability to modify human DNA could completely change how people are treated. Gene editing is a technique that can modify an organism's DNA molecularly to get a desired result. Tools like CRISPR-Cas9 are used to do this [2]. The enhancement of health outcomes and the management of disease have been the focus of gene editing research. In this regard, studies on the application of gene editing to enhance crops to increase their hardiness and tolerance to environmental challenges have been conducted [3]. There has also been study on using gene editing to change mosquito genomes to prevent the mosquitoes from spreading diseases like malaria. The studies on cattle have been conducted with the goal of modifying their genomes to make them hornless [4]. In the modern world, we can witness a revolution in the human health care system. Genome editing has become much more common in all of these disciplines since the discovery and subsequent development of CRISPR. All that research has not, however, drawn as much interest or funding from the public as has the study of possible uses of gene editing technology to address problems with human health [5].

There were differing opinions when He Jiankui revealed in 2018 that he had changed the genomes of two kids to make them immune to HIV infection. Although the scientific community strongly denounced his callous disregard for the children's safety, opinions among the public remained split. While many praised his attempt to slow the rising HIV rates, others expressed concern that he had subjected our most defenseless and helpless to what amounted to human experimentation. The children's genomes were altered while they were still embryonic [6].

There is great promise and possibility for treating and preventing human genetic disease. Human genome editing also has significant and novel safety and ethical problems [7]. The dangers, concerns, and regulatory approaches associated with different types of human genome editing vary greatly [8].

4.2 HUMAN GENOME EDITING

The process of changing an organism's DNA sequence at a particular location is known as genome editing. By employing a brief RNA or protein sequence that binds at a particular location, followed by the use of a DNA-cutting protein known as a protease to cut the DNA double helix at or close to the binding site, genome editing can produce precise modifications [9]. Unlike previous methods of DNA engineering, this procedure does not involve inserting DNA sequences at random into a genome. One of the numerous internal DNA repair mechanisms in the cell fixes the broken DNA [10]. This makes it possible to change the original DNA sequence by altering one or more nucleotides, or to add or remove nucleotides (or even larger regions of DNA), depending on the situation [11]. Protease proteins with specific functions were employed by the first gene editing systems, notably zinc-finger nucleases and transcription activator-like effector nucleases (TALENs), to attach to and cut the right DNA sequence [12]. A lot of effort and skill are required to build proteins which would bind to and cut a particular sequence of DNA. CRISPR, which is a most recent gene editing technology, uses a guide RNA molecule rather than a protein to bind to the target DNA sequence, which is combined with a DNA cutting protease such as Cas9 [13]. CRISPR technology evolved in bacteria as a quasi-immune system to cut and destroy the DNA invading bacteriophages. CRISPR/Cas9 increases the ease, precision, affordability, and utility of genome editing [14]. This is partly due to the fact that a guide RNA can be produced for any given DNA sequence far more rapidly and readily than a protein that would bind at the intended location [15]. Additionally, it can be used to perform a wide range of genetic modifications, including combinations of several simultaneous genetic changes, point mutation edits, complete gene knockouts, and the insertion of new DNA sequences, as well as carrying out additional tasks, like diagnosing and altering gene expression [16]. Although genome editing can be applied to any kind of organism, this review focuses on human genome editing, which falls into multiple categories of applications [17]. Treating the somatic cells of individuals who already have genetic diseases is the least contentious use in human medicine [18]. Government-approved clinical studies utilizing somatic gene editing are underway in nations such as the United States, China, and Europe to treat patients with cancer, sickle cell disease, and, among other things, hereditary eye disorders [19].

Germline or heritable gene editing is a far more contentious use of genome editing in humans. To alter every cell in the ensuing child, this application would entail genome editing of a fertilized egg, early embryo, or cells that will mature into gametes [20]. As will be covered later, heritable gene editing is not currently permitted in any jurisdiction. To pass on HIV resistance, a renegade Chinese scientist named He Jiankui startled the world in November 2018 by revealing that he had gene edited at least two human embryos that produced live births; the birth of a third gene-edited child was subsequently confirmed [21]. All agreed that these cases of heritable gene editing were unethical and against many national and international laws and conventions. Dr. He lost his employment, was fined, placed under house arrest, and ultimately received a three-year jail sentence from Chinese authorities [22].

The most known human applications are somatic therapy and heritable editing, but there are also other potential human applications to note. One is for nontherapeutic enhancement purposes, such as gene editing an existing person or a future person, that is, heritable editing for better abilities, improved performance, and other qualities [23].

4.3 REGULATORY ASPECTS OF HUMAN GENOME EDITING

Treating pre-existing diseases with somatic genome editing is the least contentious use for humans. Somatic gene therapy, which has been approved for hundreds of clinical trials globally, is quite comparable to this use [24]. Because genome editing entails more precise genome editing rather than, say, the random insertion of DNA segments into the genome, as occurs with many conventional gene therapy procedures, it is probably a safer process than traditional gene therapy [2]. As a result, several jurisdictions have granted approval for somatic genome editing initiatives using the same regulatory authority that has been applied to clinical trials for gene therapy [25]. Even though somatic gene editing may be and is being added to current national regulatory systems, there are still certain special safety and ethical issues with this use [26]. Because gene editing is not perfect, mistakes could happen at the location of the change, called on-site errors, or as a result of accidental alterations made to other parts of the genome, called off-site errors [27]. Depending on the target tissue, somatic gene editing can be approached in two main ways. Certain tissue types, including blood cells, can be taken out of the body and edited *in vivo* [28]. The altered cells can then be cultured in a lab setting and examined for editing faults and correctness before being put back into the body [29]. As a result, there is less control over the editing process because it is difficult to identify editing errors, and the body cannot eliminate the impacted cells [30].

Consent is one area that is still a little unclear in the national regulatory systems, which generally have comparable and well-established premarket approval procedures for somatic gene editing [31]. Genome editing is an exotic new technique, and its supporters run the risk of overstating its potential or patients developing irrational expectations about it, which could skew the informed consent process [32]. Given the unique nature of this technology, the open question is whether further consent of any kind or degree is necessary for gene editing, as well as consent for longer-term monitoring of patients with genome editing for unanticipated effects [33].

4.4 HERITABLE GENOME EDITING

The ethical issues surrounding the genetic modification of humans have been discussed for a long time [34]. Even though no country has yet legitimated heritable gene editing, national responses to it are probably going to be more diverse than those to somatic gene editing [35]. However, the recent finding of CRISPR has increased the debate's immediacy and urgency because it offers a workable technological method for making these changes [36]. This means that germline or heritable gene editing is the main area of concern when it comes to CRISPR human genome editing.

Heritable gene editing poses further safety and ethical concerns in comparison to somatic editing. Every cell in a heritable edited progeny will have the gene edit, assuming inadequate editing of only some of the cells in a multicellular embryo does not lead to mosaicism [37]. Because the entire person is subjected to the gene edit, any adverse effects will be more intense than they would be in a somatic editing scenario, in which only a small percentage of the person's cells are changed [38]. Moreover, a person who undergoes heritable editing will leave behind changed genes in their offspring, and it may take several generations before the negative effects of the editing are apparent, by which time dozens of people could be impacted [39]. If heritable modification of embryos can be made, it will take a while to prove that it is safe [40].

Heritable genome editing raises additional ethical questions in addition to safety ones. Eugenic problems could arise from genetic editing since it could allow parents to confer specific qualities on their children [41]. In the event that a very small number of parents used preimplantation genetic diagnosis and *in vitro* fertilization to select only healthy embryos for implantation, then heritable gene editing would only be necessary to prevent genetic disease in many parents who are known to be at risk of having children with a serious genetic disease [42]. Applications to try and add new qualities, like illness immunity or enhanced traits, are more common [43].

The ability of individuals or governments to permanently alter the human genome—which some ethical declarations assert is the shared heritage of all people—raises ethical questions in light of heritable genome editing [44]. Consent is another ethical issue, as future generations will never have the chance to give their consent to procedures like heritable genome editing that could have a significant impact on their health and welfare, but parents constantly make decisions that could have an impact on their current or future children [44].

Given the serious and unresolved issues concerned with the safety and ethics of heritable genome editing, there is widespread consensus that this is not the moment to do such editing [45]. On the question of whether heritable gene editing should be permanently forbidden, there is, nevertheless, a genuine dispute that has given rise to differing opinions [46]. Some scientists, ethicists, and countries want to outlaw human heritable gene editing forever, arguing that it is intrinsically wrong to alter the human genetic pool in this way, even if the safety concerns can be resolved [47].

On the other hand, some argue that human genetic editing can be ethically justified and should be allowed to move forward, provided that safety issues are addressed. For example, two separate committees established by the US National Academy of Sciences have recommended that if human heritable editing is to proceed, it should be limited at least initially to restoring wild-type (normal) function in future children with serious, monogenic genetic illnesses [48]. Although heritable gene editing has not yet been approved by any country, national regulatory frameworks handle the practice in a more or less limited manner [49]. It is challenging to distinguish between national regulatory restrictions on heritable editing because various nations have previously imposed limitations on related practices such as *in vitro* reproduction, human reproductive cloning, stem cell applications, and heritable gene therapy that may or may not also apply to heritable gene editing, depending on how the measure is worded [50].

There are more permissive regulatory frameworks that do not specifically forbid heritable gene editing; but, in order to move further, specific requests would need to be approved, and none of these have been [51]. Nonetheless, these nations allow studies and tests that do not result in the implantation of altered embryos and that modify gametes, embryos, or gamete precursor cells [52]. These jurisdictions include, for example, China and the United Kingdom. Even though national regulatory systems are uniform today in that they do not approve heritable gene editing attempts, there is no guarantee that international harmonization will continue in the future given the various regulatory approaches of various countries, which likely reflect divergent social, political, historical, and ethical norms in each nation [53].

4.5 ENHANCEMENT

Human genome editing has primarily been used for health-related purposes; it may also be used in the future for non-health-related purposes, such as genetic enhancement. The possibility of genetic editing improving human performance raises a number of concerns [54]. The topic of genetic enhancement has long been discussed, but, similar to heritable alteration, the development of human genome editing gives the discussion more urgency and substance [55].

Most of the discussions about genetic enhancement are on characteristics like intellect, physical prowess, looks, or disposition. But during the last 20 years, genetic research has shown that the complicated etiologies of these features entail many genes working in concert with the environment [56]. Because of this, it will be challenging to significantly improve these important features by changing one or even a few genes. Competitive sports are the most likely use for human gene editing for enhancement [57]. Professional and Olympic athletes have a long history of utilizing illegal drugs to try to get an advantage over their competitors. The World Anti-Doping Agency has previously ruled gene editing for the purpose of sports enhancement to be unlawful, and the same motivations would apply in this case. There are genes known to affect athletic performance, and even if a single gene change has little effect, people will probably try anything to get a competitive advantage, legal or illegal [58]. Risk-benefit balance is a subject that is related to feasibility. It is possible for genetic enhancement to take place in heritable or somatic applications. Like gene editing for therapeutic purposes, gene editing would involve risks of unintended effects, with the risks being greater in heritable than somatic applications, but with even somatic editing bringing nonnegligible concerns [59]. Given the potentially life-saving benefits of therapeutic gene editing applications in the treatment or prevention of serious diseases, the risks may occasionally be sufficiently reduced and controlled to make the approach acceptable. However, the gains for improvement applications are likely to be far less predictable and negligible, making them less worthwhile to take any risks for [60]. Furthermore, although public opinion is more supportive of enhancement in some Asian countries, it is particularly strong against the use of gene editing for enhancement in many other countries.

The challenge of distinctly differentiating between a medical intervention's therapeutic and enhancement purposes is another significant enhancement-related concern. Numerous conditions exhibit a severity spectrum [61]. Targeting the condition at one extreme, where the individual may be adversely affected physically or psychologically due to abnormally low or high measurements, could be considered therapeutic [62]. At the other end of the spectrum, where the individual falls comfortably within the normal range of variation, such interventions could be considered enhancement [63]. Furthermore, social constructs and contextual factors influence how diseases are defined and experienced, so a state that is considered pathological in one setting may not be in another [64].

Although no nation has allowed or authorized the use of human gene editing for improvement, many jurisdictions lack the necessary regulatory frameworks. The European signatories to the Oviedo Convention, which forbids genetic enhancement, are one exception [65]. A few other nations have outright banned genetic enhancement (Chile, Columbia, Mexico, and Panama, to name a few). While many Asian nations tend to be more accepting of human enhancement, none have made it clear that they endorse genome editing enhancement. Otherwise, the majority of current regulatory frameworks concentrate on approving goods that are used in illness diagnosis, prevention, or treatment [66]. The compatibility of an enhancement technology with these current regulatory frameworks is unclear.

4.6 BASIC AND APPLIED RESEARCH

Research is the fourth way that human genome editing is being used. Somatic cell editing research is generally not controversial and does not present any special ethical or regulatory issues [67]. The regulation of research involving embryos or gametes is subject to controversy and varies among nations due to pre-existing legislative limitations [68]. Research on human embryos used for genome editing may be conducted to better understand the fundamentals of human reproduction, the development of early embryos, and the course of different genetic illnesses, or it may be done to prepare for potential future heritable modification [69].

Human genome editing of human embryos for research purposes is allowed in a number of countries, including China, the US, the UK, and Sweden. For instance, research from the US and China has sought to use CRISPR to treat genetic disorders in human embryos that are viable or not [70]. Embryonic genome editing studies in the United Kingdom have focused on investigating the reproductive process and possible ways for solving human fertility. Human embryo research is prohibited in some nations, and this prohibition also extends to any experiments using genome editing on human embryos. Germany, France, and Italy are a few examples [71].

The lack of consensus on definitions for certain terminology, such as "human embryo" and "germline," adds to the complexity of the disparities across national laws governing human embryo genome editing research [72]. Regulations allowing human editing in embryo research are unlikely to be entirely harmonized throughout nations due to the variety in national laws, which are firmly ingrained in each country's history and culture [73]. However, there might be other goals for international coordination, such as assembling a group of countries that can move forward with embryonic research and sharing resources, techniques, and outcomes with countries that aren't allowed to do it themselves [74].

4.7 INTERNATIONAL GOVERNANCE

Numerous proposals for international administration of this technology have been made in response to differing state viewpoints and open-ended questions regarding oversight of human genome editing [75]. The first topic covered in this section is the necessity, justification, and method of international governance of human gene editing. It then goes over the current international governance framework for human gene editing [76]. The last part covers new efforts or methods for international governance that have been proposed [77].

4.8 SCOPE OF LAW

International methods for human genome editing governance have been called for on numerous occasions [78]. However, there are grounds against international government; thus, its case is not unavoidable. Regarding human genome editing, each nation has unique cultural, political, economic, and ethical values that are relevant [79]. These values may support varying regulatory approaches and preferences. Furthermore, negotiating international governance agreements requires a significant investment of time, energy, and financial resources that would be better used for domestic regulation, particularly in less developed countries [80]. Even in cases where negotiations for an international agreement are successful, their compliance rates are typically low because there is typically no enforcement body with the necessary power [81].

The first justification for international control over human gene editing is the defense of medical ethics and universal human rights [82]. Although political, economic, and legal systems are mostly left to the sovereignty of individual nations, violations of basic human rights or significant medical ethical precepts affect all nations and peoples [83]. Everybody is concerned when someone is tortured, denied their political and fundamental liberties, or put through dangerous and unethical medical procedures like experimental genome editing [84].

Second, when the resource being safeguarded is the shared property of all people, international technology governance is justified [85]. The most famous examples come from the global ecosystem, such as the stratospheric ozone layer or climate change, where all countries are entitled to a shared ownership of the resource and are required to make contributions toward its protection [86]. One may make a similar case for the human gene pool, arguing that it belongs to all of humanity and that any attempt to permanently change it should be decided upon by all nations and peoples [87].

Fairness is the third justification for international control over human genome editing [88]. The use of genome editing for enhancement in sports could give certain countries an unfair competitive advantage over others if such use is permitted in some countries but not in others [89]. Such unfairness may be able to be identified, prevented, and enforced against by stronger international governance institutions [90].

Lastly, there's a pragmatic case for global government. Global discussions offer chances to gather and exchange thoughts, viewpoints, experiences, and outcomes from all over the world [91]. These pooled resources can weave together a shared vision for the future of human genome editing, while also decreasing the need for every country to perform all the work itself and in a duplicative way [92].

4.9 CONCLUSION

Human genome editing presents the potential to treat severe sickness as well as curing severe sickness from the individual's family tree. There will always be conflict between the interests of the general public's health and the scientific community's desire to advance research into commercially viable treatments if it is left unregulated. A rights-based approach to science will guarantee that scientific advancement may continue within legal bounds and that patients' and clinical trial participants' legal rights are safeguarded. Properly regulated, the possibility of unethical research will be minimized.

4.10 FUTURE SCOPE

Public access to the clinical uses of heritable human genome editing should only occur if these uses have been demonstrated to be secure and efficient. Heritable human genome editing that is not therapeutic should be governed similarly to heritable human genome editing, with the caveat that it cannot, in any legally significant sense, cause harm.

KEYWORDS

- **biomedical research**
- **CRISPR**
- **gene**
- **genetic disease**
- **genome editing**
- **germline**
- **human embryo**
- **plant breeding**

REFERENCES

1. Alex, K., & Winkler, E. C. (2024). Comparative ethical evaluation of epigenome editing and genome editing in medicine: First steps and future directions. *Journal of Medical Ethics, 50*(6), 398–406.
2. Aljabali, A. A., El-Tanani, M., & Tambuwala, M. M. (2024). Principles of CRISPR-Cas9 technology: Advancements in genome editing and emerging trends in drug delivery. *Journal of Drug Delivery Science and Technology*, 105338.
3. Pacesa, M., Pelea, O., & Jinek, M. (2024). Past, present, and future of CRISPR genome editing technologies. *Cell, 187*(5), 1076–1100.
4. Irfan, M., Majeed, H., Iftikhar, T., & Ravi, P. K. (2024). A review on molecular scissoring with CRISPR/Cas9 genome editing technology. *Toxicology Research, 13*(4).
5. Khoshandam, M., Soltaninejad, H., Mousazadeh, M., Hamidieh, A. A., & Hosseinkhani, S. (2024). Clinical applications of the CRISPR/Cas9 genome-editing system: Delivery options and challenges in precision medicine. *Genes & Diseases, 11*(1), 268–282.
6. Nelson, J. P., Tomblin, D. C., Barbera, A., & Smallwood, M. (2024). The divide so wide: Public perspectives on the role of human genome editing in the US healthcare system. *Public Understanding of Science, 33*(2), 189–209.
7. Mesaki, K., Yamamoto, H., Juvet, S., Yeung, J., Guan, Z., Akhter, A., Yao, Y., Dickie, C., Mangat, H., Wang, A., & Wilson, G. W. (2024). CRISPR-Cas genome editing in *ex vivo* human lungs to rewire the translational path of genome-targeting therapeutics. *Human Gene Therapy.*
8. Callegari, A., Kueblbeck, M., Morero, N. R., Serrano-Solano, B., & Ellenberg, J. (2024). Rapid generation of homozygous fluorescent knock-in human cells using CRISPR–Cas9 genome editing and validation by automated imaging and digital PCR screening. *Nature Protocols*, 1–41.
9. Morshedzadeh, F., Ghanei, M., Lotfi, M., Ghasemi, M., Ahmadi, M., Najari-Hanjani, P., Sharif, S., Mozaffari-Jovin, S., Peymani, M., & Abbaszadegan, M. R. (2024). An update on the application of CRISPR technology in clinical practice. *Molecular Biotechnology, 66*(2), 179–197.
10. Parums, D. V. (2024). Genome editing goes beyond CRISPR with the emergence of 'Bridge' RNA editing. *Medical Science Monitor: International Medical Journal of Experimental and Clinical Research, 30*, e945933-1.
11. Philippidis, A. (2024). CASGEVY makes history as FDA approves first CRISPR/Cas9 genome edited therapy. *Human Gene Therapy, 35*(1, 2), 1–4.
12. Hu, C., Myers, M. T., Zhou, X., Hou, Z., Lozen, M. L., Nam, K. H., Zhang, Y., & Ke, A. (2024). Exploiting activation and inactivation mechanisms in type IC CRISPR-Cas3 for genome-editing applications. *Molecular Cell, 84*(3), 463–475.
13. Gupte, P., Dhingra, K., & Saloni, V. (2024). Precision gene editing strategies with CRISPR-Cas9 for advancing cancer immunotherapy and Alzheimer's disease. *Journal of Knowledge Learning and Science Technology, 3*(4), 11–21.
14. Parums, D. V. (2024). First regulatory approvals for CRISPR-Cas9 therapeutic gene editing for sickle cell disease and transfusion-dependent β-thalassemia. *Medical*

Science Monitor: International Medical Journal of Experimental and Clinical Research, 30, e944204-1.
15. Singh, A., Smedley, G. D., Rose, J. G., Fredriksen, K., Zhang, Y., Li, L., & Yuan, S. H. (2024). A high efficiency precision genome editing method with CRISPR in iPSCs. *Scientific Reports, 14*(1), 9933.
16. Chehelgerdi, M., Chehelgerdi, M., Khorramian-Ghahfarokhi, M., Shafieizadeh, M., Mahmoudi, E., Eskandari, F., Rashidi, M., Arshi, A., & Mokhtari-Farsani, A. (2024). Comprehensive review of CRISPR-based gene editing: Mechanisms, challenges, and applications in cancer therapy. *Molecular Cancer, 23*(1), 9.
17. Huang, K., Qu, Y., Cousins, H., Johnson, W. A., Yin, D., Shah, M., Zhou, D., Altman, R., Wang, M., & Cong, L. (2024). Crispr-GPT: An LLM agent for automated design of gene-editing experiments. *arXiv preprint arXiv:2404.18021.*
18. Capin, J., Harrison, A., Raele, R. A., Yadav, S. K., Baiwir, D., Mazzucchelli, G., Quinton, L., Satchwell, T. J., Toye, A. M., Schaffitzel, C., & Berger, I. (2024). An engineered baculoviral protein and DNA co-delivery system for CRISPR-based mammalian genome editing. *Nucleic Acids Research, 52*(6), 3450–3468.
19. Qin, M., Deng, C., Wen, L., Luo, G., & Meng, Y. (2024). CRISPR-Cas and CRISPR-based screening system for precise gene editing and targeted cancer therapy. *Journal of Translational Medicine, 22*(1), 516.
20. Badon, I. W., Oh, Y., Kim, H. J., & Lee, S. H. (2024). Recent application of CRISPR-Cas12 and OMEGA system for genome editing. *Molecular Therapy, 32*(1), 32–43.
21. Karpov, D. S. (2024). CRISPR-Cas systems and genome editing: Beginning the era of CRISPR/Cas therapies for humans. *International Journal of Molecular Sciences, 25*(10), 5292.
22. Lopes, R., & Prasad, M. K. (2024). Beyond the promise: Evaluating and mitigating off-target effects in CRISPR gene editing for safer therapeutics. *Frontiers in Bioengineering and Biotechnology, 11*, 1339189.
23. Liang, R., He, Z., Zhao, K. T., Zhu, H., Hu, J., Liu, G., Gao, Q., Liu, M., Zhang, R., Qiu, J. L., & Gao, C. (2024). Prime editing using CRISPR-Cas12a and circular RNAs in human cells. *Nature Biotechnology*, 1–9.
24. Vasconcelos Komninakis, S., Domingues, W., Saeed Sanabani, S., Angelo Folgosi, V., Neves Barbosa, I., & Casseb, J. (2024). CRISPR/Cas as a powerful tool for human immunodeficiency virus cure: A review. *AIDS Research and Human Retroviruses.*
25. Muhsin, S. M., binti Awang, A. F., & Boon, A. H. (2024). Ethical considerations in human genome editing: Exploring CRISPR technology.
26. Cho, H. Y., Yoo, M., Pongkulapa, T., Rabie, H., Muotri, A. R., Yin, P. T., Choi, J. W., & Lee, K. B. (2024). Magnetic nanoparticle-assisted non-viral CRISPR-Cas9 for enhanced genome editing to treat Rett syndrome. *Advanced Science*, 2306432.
27. Corrado, A., Aceto, R., Miglietta, S., Silvestri, R., Dell'Anno, I., Lepori, I., Ricci, B., Romei, C., Giovannoni, R., Poliseno, L., & Evangelista, M. (2024). Strategies for single base gene editing in an immortalized human cell line by CRISPR/Cas9 technology. *3 Biotech, 14*(2), 45.
28. Dubey, S., Chen, Z., Jiang, Y. J., Talis, A., Molotkov, A., Ali, A., Mintz, A., & Momen-Heravi, F. (2024). Small extracellular vesicles (sEVs)-based gene delivery platform for cell-specific CRISPR/Cas9 genome editing. *Theranostics, 14*(7), 2777.

29. Zhu, M., Xu, R., Yuan, J., Wang, J., Ren, X., Cong, T., You, Y., Ju, A., Xu, L., Wang, H., & Zheng, P. (2024). Tracking-seq reveals the heterogeneity of off-target effects in CRISPR–Cas9-mediated genome editing. *Nature Biotechnology*, 1–2.
30. Villiger, L., Joung, J., Koblan, L., Weissman, J., Abudayyeh, O. O., & Gootenberg, J. S. (2024). CRISPR technologies for genome, epigenome and transcriptome editing. *Nature Reviews Molecular Cell Biology*, 1–24.
31. Vlashi, R., Zhang, X., & Li, H., & Chen, G. (2024). Potential therapeutic strategies for osteoarthritis via CRISPR/Cas9-mediated gene editing. *Reviews in Endocrine and Metabolic Disorders, 25*(2), 339–367.
32. Asmamaw Mengstie, M., Teshome Azezew, M., Asmamaw Dejenie, T., Teshome, A. A., Tadele Admasu, F., Behaile Teklemariam, A., Tilahun Mulu, A., Mekonnen Agidew, M., Adugna, D. G., Geremew, H., & Abebe, E. C. (2024). Recent advancements in reducing the off-target effect of CRISPR-Cas9 genome editing. *Biologics: Targets and Therapy*, 21–28.
33. Wen, B., Li, E., Wang, G., Kalin, T. R., Gao, D., Lu, P., Kalin, T. V., & Kalinichenko, V. V. (2024). CRISPR-Cas9 genome editing allows generation of the mouse lung in a rat. *American Journal of Respiratory and Critical Care Medicine, 210*(2), 167–177.
34. Han, A. R., Shin, H. R., Kweon, J., Lee, S. B., Lee, S. E., Kim, E. Y., Chang, E. J., Kim, Y., & Kim, S. W. (2024). Highly efficient genome editing via CRISPR-Cas9 ribonucleoprotein (RNP) delivery in mesenchymal stem cells. *BMB Reports, 57*(1), 60.
35. Li, T., Li, S., Kang, Y., Zhou, J., & Yi, M. (2024). Harnessing the evolving CRISPR/Cas9 for precision oncology. *Journal of Translational Medicine, 22*(1), 749.
36. Ponta, S., Bonato, A., Neidenbach, P., Bruhin, V. F., Laurent, A., Applegate, L. A., Zenobi-Wong, M., & Barreto, G. (2024). Streamlined, single-step non-viral CRISPR-Cas9 knockout strategy enhances gene editing efficiency in primary human chondrocyte populations. *Arthritis Research & Therapy, 26*(1), 66.
37. Bevacqua, R. J., Zhao, W., Merheb, E., Kim, S. H., Marson, A., Gloyn, A. L., & Kim, S. K. (2024). Multiplexed CRISPR gene editing in primary human islet cells with Cas9 ribonucleoprotein. *iScience, 27*(1).
38. Allemailem, K. S., Almatroudi, A., Rahmani, A. H., Alrumaihi, F., Alradhi, A. E., Alsubaiyel, A. M., Algahtani, M., Almousa, R. M., Mahzari, A., Sindi, A. A., & Dobie, G. (2024). Recent updates of the CRISPR/Cas9 genome editing system: Novel approaches to regulate its spatiotemporal control by genetic and physicochemical strategies. *International Journal of Nanomedicine*, 5335–5363.
39. Ruffolo, J. A., Nayfach, S., Gallagher, J., Bhatnagar, A., Beazer, J., Hussain, R., Russ, J., Yip, J., Hill, E., Pacesa, M., & Meeske, A. J. (2024). Design of highly functional genome editors by modeling the universe of CRISPR-Cas sequences. *bioRxiv*.
40. Atia, M., Jiang, W., Sedeek, K., Butt, H., & Mahfouz, M. (2024). Crop bioengineering via gene editing: Reshaping the future of agriculture. *Plant Cell Reports, 43*(4), 98.
41. Klermund, J., Rhiel, M., Kocher, T., Chmielewski, K. O., Bischof, J., Andrieux, G., El Gaz, M., Hainzl, S., Boerries, M., Cornu, T. I., & Koller, U. (2024). On- and off-target effects of paired CRISPR-Cas nickase in primary human cells. *Molecular Therapy, 32*(5), 1298–1310.
42. Zhang, Z., Baxter, A. E., Ren, D., Qin, K., Chen, Z., Collins, S. M., Huang, H., Komar, C. A., Bailer, P. F., Parker, J. B., & Blobel, G. A. (2024). Efficient engineering

of human and mouse primary cells using peptide-assisted genome editing. *Nature Biotechnology, 42*(2), 305–315.
43. Abkallo, H. M., Arbuthnot, P., Auer, T. O., Berger, D. K., Burger, J., Chakauya, E., Concordet, J. P., Diabate, A., Di Donato, V., Groenewald, J. H., & Guindo, A. (2024). Making genome editing a success story in Africa. *Nature Biotechnology, 42*(4), 551–554.
44. Tong, L. W., Hu, Y. S., Yu, S. J., Li, C. L., & Shao, J. W. (2024). Current application and future perspective of CRISPR/Cas9 gene editing system-mediated immune checkpoint for liver cancer treatment. *Nanotechnology, 35*(40), 402002.
45. Wang, Y., Lin, R. Z., Harris, M., Lavayen, B., McCreedy, B., Hofmeister, R., & Getts, D. (2024). CRISPR-enabled autonomous transposable element (CREATE) for RNA-based gene editing and delivery. *bioRxiv.* https://doi.org/10.1101/2024.01.30.000000 (accessed on 5th January 2026).
46. Meng, X., Jia, R., Zhao, X., Zhang, F., Chen, S., Yu, S., Liu, X., Dou, H., Feng, X., Zhang, J., & Wang, N. (2024). *In vivo* genome editing via CRISPR/Cas9-mediated homology-independent targeted integration for Bietti crystalline corneoretinal dystrophy treatment. *Nature Communications, 15*(1), 3773.
47. He, Y., Han, Y., Ma, Y., Liu, S., Fan, T., Liang, Y., Tang, X., Zheng, X., Wu, Y., Zhang, T., & Qi, Y. (2024). Expanding plant genome editing scope and profiles with CRISPR-FrCas9 systems targeting palindromic TA sites. *Plant Biotechnology Journal.* https://doi.org/10.1111/pbi.14157 (accessed on 5th January 2026).
48. Acharya, S., Ansari, A. H., Kumar Das, P., Hirano, S., Aich, M., Rauthan, R., Mahato, S., Maddileti, S., Sarkar, S., Kumar, M., & Phutela, R. (2024). PAM-flexible engineered FnCas9 variants for robust and ultra-precise genome editing and diagnostics. *Nature Communications, 15*(1), 5471.
49. Macarrón Palacios, A., Korus, P., Wilkens, B. G., Heshmatpour, N., & Patnaik, S. R. (2024). Revolutionizing *in vivo* therapy with CRISPR/Cas genome editing: Breakthroughs, opportunities, and challenges. *Frontiers in Genome Editing, 6*, 1342193.
50. Dixit, S., Kumar, A., Srinivasan, K., Vincent, P. D., & Ramu Krishnan, N. (2024). Advancing genome editing with artificial intelligence: Opportunities, challenges, and future directions. *Frontiers in Bioengineering and Biotechnology, 11*, 1335901.
51. Khalil, A. M. (2020). The genome editing revolution. *Journal of Genetic Engineering and Biotechnology, 18*(1), 68.
52. Wang, J. Y., & Doudna, J. A. (2023). CRISPR technology: A decade of genome editing is only the beginning. *Science, 379*(6629), eadd8643.
53. Doudna, J. A. (2020). The promise and challenge of therapeutic genome editing. *Nature, 578*(7794), 229–236.
54. Slaymaker, I. M., & Gaudelli, N. M. (2021). Engineering Cas9 for human genome editing. *Current Opinion in Structural Biology, 69*, 86–98.
55. Alagoz, M., & Kherad, N. (2020). Advanced genome editing technologies in the treatment of human diseases: CRISPR therapy. *International Journal of Molecular Medicine, 46*(2), 521–534.
56. Najafi, S., Tan, S. C., Aghamiri, S., Raee, P., Ebrahimi, Z., Jahromi, Z. K., Rahmati, Y., Nahand, J. S., Piroozmand, A., Jajarmi, V., & Mirzaei, H. (2022). Therapeutic

potentials of CRISPR-Cas genome editing technology in human viral infections. *Biomedicine & Pharmacotherapy, 148*, 112743.
57. Nambiar, T. S., Baudrier, L., Billon, P., & Ciccia, A. (2022). CRISPR-based genome editing through the lens of DNA repair. *Molecular Cell, 82*(2), 348–388.
58. Xue, C., & Greene, E. C. (2021). DNA repair pathway choices in CRISPR-Cas9-mediated genome editing. *Trends in Genetics, 37*(7), 639–656.
59. Li, Y., Glass, Z., Huang, M., Chen, Z. Y., & Xu, Q. (2020). *Ex vivo* cell-based CRISPR/Cas9 genome editing for therapeutic applications. *Biomaterials, 234*, 119711.
60. Sledzinski, P., Dabrowska, M., Nowaczyk, M., & Olejniczak, M. (2021). Paving the way towards precise and safe CRISPR genome editing. *Biotechnology Advances, 49*, 107737.
61. Bevacqua, R. J., Dai, X., Lam, J. Y., Gu, X., Friedlander, M. S., Tellez, K., Miguel-Escalada, I., Bonàs-Guarch, S., Atla, G., Zhao, W., & Kim, S. H. (2021). CRISPR-based genome editing in primary human pancreatic islet cells. *Nature Communications, 12*, 2397.
62. Barman, A., Deb, B., & Chakraborty, S. (2020). A glance at genome editing with CRISPR–Cas9 technology. *Current Genetics, 66*, 447–462.
63. Hirakawa, M. P., Krishnakumar, R., Timlin, J. A., Carney, J. P., & Butler, K. S. (2020). Gene editing and CRISPR in the clinic: Current and future perspectives. *Bioscience Reports, 40*(4), BSR20200127.
64. Bhattacharjee, G., Mani, I., Gohil, N., Khambhati, K., Braddick, D., Panchasara, H., & Singh, V. (2020). CRISPR technology for genome editing. In *Precision Medicine for Investigators, Practitioners and Providers* (pp. 59–69). Academic Press.
65. Taha, E. A., Lee, J., & Hotta, A. (2022). Delivery of CRISPR-Cas tools for *in vivo* genome editing therapy: Trends and challenges. *Journal of Controlled Release, 342*, 345–361.
66. Li, H., Yang, Y., Hong, W., Huang, M., Wu, M., & Zhao, X. (2020). Applications of genome editing technology in the targeted therapy of human diseases: Mechanisms, advances and prospects. *Signal Transduction and Targeted Therapy, 5*(1), 1.
67. Doxzen, K., & Halpern, J. (2020). Focusing on human rights: A framework for CRISPR germline genome editing ethics and regulation. *Perspectives in Biology and Medicine, 63*(1), 44–53.
68. Karapurkar, J. K., Antao, A. M., Kim, K. S., & Ramakrishna, S. (2021). CRISPR-Cas9 based genome editing for defective gene correction in humans and other mammals. *Progress in Molecular Biology and Translational Science, 181*, 185–229.
69. Lim, J. M., & Kim, H. H. (2022). Basic principles and clinical applications of CRISPR-based genome editing. *Yonsei Medical Journal, 63*(2), 105.
70. Arroyo-Olarte, R. D., Bravo Rodriguez, R., & Morales-Ríos, E. (2021). Genome editing in bacteria: CRISPR-Cas and beyond. *Microorganisms, 9*(4), 844.
71. Kan, M. J., & Doudna, J. A. (2022). Treatment of genetic diseases with CRISPR genome editing. *JAMA, 328*(10), 980–981.
72. El-Mounadi, K., Morales-Floriano, M. L., & Garcia-Ruiz, H. (2020). Principles, applications, and biosafety of plant genome editing using CRISPR-Cas9. *Frontiers in Plant Science, 11*, 56.

73. Quinn, J., Musa, A., Kantor, A., McClements, M. E., Cehajic-Kapetanovic, J., MacLaren, R. E., & Xue, K. (2021). Genome-editing strategies for treating human retinal degenerations. *Human Gene Therapy, 32*(5, 6), 247–259.
74. Ebrahimi, V., & Hashemi, A. (2020). Challenges of *in vitro* genome editing with CRISPR/Cas9 and possible solutions: A review. *Gene, 753*, 144813.
75. Denes, C. E., Cole, A. J., Aksoy, Y. A., Li, G., Neely, G. G., & Hesselson, D. (2021). Approaches to enhance precise CRISPR/Cas9-mediated genome editing. *International Journal of Molecular Sciences, 22*(16), 8571.
76. Zhang, D., Hussain, A., Manghwar, H., Xie, K., Xie, S., Zhao, S., Larkin, R. M., Qing, P., Jin, S., & Ding, F. (2020). Genome editing with the CRISPR-Cas system: An art, ethics and global regulatory perspective. *Plant Biotechnology Journal, 18*(8), 1651–1669.
77. Rees, H. A., Minella, A. C., Burnett, C. A., Komor, A. C., & Gaudelli, N. M. (2021). CRISPR-derived genome editing therapies: Progress from bench to bedside. *Molecular Therapy, 29*(11), 3125–3139.
78. Asmamaw, M., & Zawdie, B. (2021). Mechanism and applications of CRISPR/Cas-9-mediated genome editing. *Biologics: Targets and Therapy*, *15*, 353–361.
79. Akram, F., Sahreen, S., Aamir, F., Haq, I. U., Malik, K., Imtiaz, M., Naseem, W., Nasir, N., & Waheed, H. M. (2023). An insight into modern targeted genome-editing technologies with a special focus on CRISPR/Cas9 and its applications. *Molecular Biotechnology, 65*(2), 227–242.
80. Naeem, M., Majeed, S., Hoque, M. Z., & Ahmad, I. (2020). Latest developed strategies to minimize the off-target effects in CRISPR-Cas-mediated genome editing. *Cells, 9*(7), 1608.
81. Tyagi, S., Kumar, R., Das, A., Won, S. Y., & Shukla, P. (2020). CRISPR-Cas9 system: A genome-editing tool with endless possibilities. *Journal of Biotechnology, 319*, 36–53.
82. Guo, N., Liu, J. B., Li, W., Ma, Y. S., & Fu, D. (2022). The power and the promise of CRISPR/Cas9 genome editing for clinical application with gene therapy. *Journal of Advanced Research, 40*, 135–152.
83. Van der Oost, J., & Patinios, C. (2023). The genome editing revolution. *Trends in Biotechnology, 41*(3), 396–409.
84. Zheng, L., Tan, Y., Hu, Y., Shen, J., Qu, Z., Chen, X., Ho, C. L., Leung, E. L., Zhao, W., & Dai, L. (2022). CRISPR/Cas-based genome editing for human gut commensal *Bacteroides* species. *ACS Synthetic Biology, 11*(1), 464–472.
85. Mushtaq, M., Ahmad Dar, A., Skalicky, M., Tyagi, A., Bhagat, N., Basu, U., Bhat, B. A., Zaid, A., Ali, S., Dar, T. U., & Rai, G. K. (2021). CRISPR-based genome editing tools: Insights into technological breakthroughs and future challenges. *Genes, 12*(6), 797.
86. Li, Z. H., Wang, J., Xu, J. P., Wang, J., & Yang, X. (2023). Recent advances in CRISPR-based genome editing technology and its applications in cardiovascular research. *Military Medical Research, 10*(1), 12.
87. Hurlbut, J. B. (2020). Imperatives of governance: Human genome editing and the problem of progress. *Perspectives in Biology and Medicine, 63*(1), 177–194.

88. Anzalone, A. V., Koblan, L. W., & Liu, D. R. (2020). Genome editing with CRISPR–Cas nucleases, base editors, transposases and prime editors. *Nature Biotechnology, 38*(7), 824–844.
89. Hazafa, A., Mumtaz, M., Farooq, M. F., Bilal, S., Chaudhry, S. N., Firdous, M., Naeem, H., Ullah, M. O., Yameen, M., Mukhtiar, M. S., & Zafar, F. (2020). CRISPR/Cas9: A powerful genome editing technique for the treatment of cancer cells with present challenges and future directions. *Life Sciences, 263*, 118525.
90. Barman, N. C., Khan, N. M., Islam, M., Nain, Z., Roy, R. K., Haque, A., & Barman, S. K. (2020). CRISPR-Cas9: A promising genome editing therapeutic tool for Alzheimer's disease—A narrative review. *Neurology and Therapy, 9*, 419–434.
91. Doxzen, K. (2023). *Editing humanity: The CRISPR revolution and the new era of genome editing. Jurimetrics, 63*(2), 197–211.
92. Adashi, E. Y., Burgess, M. M., Burall, S., Cohen, I. G., Fleck, L. M., Harris, J., Holm, S., Lafont, C., Moreno, J. D., Neblo, M. A., & Niemeyer, S. J. (2020). Heritable human genome editing: The public engagement imperative. *The CRISPR Journal, 3*(6), 434–439.

CHAPTER 5

Precision Medicine Through Gene Editing: Pharmacogenomics and the Future of Disease Management in Clinical Healthcare

KOZMA DOROTTYA EDINA,[1*] MANMEET KAUR ARORA,[2*] BHUPINDER SINGH,[3*] and SAHIL LAL[4*]

[1]*University of Pannonia, Veszprém, Hungary*

[2]*Sharda University, Greater Noida, Uttar Pradesh, India*

[3]*Sharda School of Law, Sharda University, Greater Noida, India; University of South Wales, UK; Santo Tomas University Tunja, Colombia*

[4]*School of Law, Galgotias University, India*

*Corresponding authors

ABSTRACT

In recent years, precision medicine has been hailed as a revolution in healthcare due to its successful transition from mass treatment efforts to individualized approaches based on genomic and molecular profiling. This chapter discusses the promise of gene editing coupled with pharmacogenomics, the twin pillars of precision medicine, to improve disease management. Emerging synergistic strategies integrating these technologies yield enhanced therapeutic efficacy and reduced adverse events for a range of diseases, including oncology, genetic disorders, cardiovascular diseases, and infectious diseases. However, ethical, social, and regulatory issues relating to equity, access, and germline editing need to be addressed before responsible applications can be realized. Technologies such as high-throughput screening, big data analytics, artificial

Genetic Editing for Healthcare: Advances in Disease Management.
Bhupinder Singh, Rishabha Malviya, Sahil Lal, and Sathvik Belagodu Sridhar (Eds.)

DOI: 10.1201/9781779643827-5

intelligence, and point-of-care diagnostics are ushering in the next generation of precision medicine, with a more proactive, predictive, and personalized healthcare paradigm for a healthier tomorrow.

5.1 INTRODUCTION AND BACKGROUND

The 21st century signals a revolution in health, a new paradigm of how we view disease and a new way of going about fighting it. But now we are on the brink of precision medicine, which focuses on targeting treatments very specifically rather than generalized treatment of problems. Your journey begins in this chapter, which will lead you through the innovations that promise to define this revolution; particularly, the explosive potential of the combination of gene editing and pharmacogenomics [1]. For the most part in medical history, healthcare has been conducted according to the one-size-fits-all paradigm. This paradigm, while foundational and responsible for many of the advances we have seen in public health and disease management, typically assumes that individuals with the same diagnosis will respond the same way to any intervention. Traditionally, the practice of medicine has been reactive, consisting principally of addressing the consequences of well-established diseases according to population-level averages and clinical observations encompassing broad patient groups. Drug dosages, treatment protocols, and preventive measures tended to be standardized and were supposed to be effective for the majority, but always failed half the time for a rather significant minority.

This traditional model, though sensible in its day, is increasingly understood to be fundamentally flawed. It does not account for a beautiful mosaic of human biological diversity. Every human being is different; they have their own genetic, environmental exposure history, lifestyle, and health history. They have a significant impact on susceptibility to diseases, their progression, and response to therapeutic interventions. A drug that works miracles for one individual may do nothing or else be potentially injurious to another, even if they have exactly the same diagnosis. It is not noise in the system; it is the core of human individualism and a powerful determinant of health outcomes [2].

One-disease drug, one-shot cure, one solution to all through a single route have been the way of discovering medications but are unfortunately declining solutions for healing. With the rise of the genomic era, insights brought about by DNA sequencing and bioinformatics have revealed the deep genetic diversity present in the human population. However, our research is now showing us that even diseases that appear to follow a similar clinical path at first diagnosis—for example, certain types of cancer—often arise from complicated molecular processes involving a combination of numerous genes that are spurred by multiple environmental influences and complex gene-environment interactions. This complexity has a clear call for a more personalized, nuanced approach to healthcare [3].

It is fair to say that over the nine years since the concepts were announced, the enabling forces have converged, propelling healthcare towards personalized approaches and precision medicine from a distant dream to a real and rapidly moving reality. First, the tremendous decrease in the cost and time needed to sequence a genome has made it more accessible, if not always effectively analyzed. The Human Genome Project, which was finished in the early 2000s, was a gargantuan effort, but whole genome sequencing can now be done fairly rapidly and inexpensively [5]. Genome-wide association studies (GWAS), which began to flourish about a decade ago, identify and nurture associations between millions of genetic variations and conditions, and this bold and serendipitous playground of genetic data access serves as a foundation of precision medicine.

Secondly, developments in bioinformatics, data science, and computational biology equally provided methods to analyze and interpret the big data arising from sequencing and other omics (transcriptomics, proteomics, and metabolomics) technologies. But now, within this complex system, the ability to access these "big data" approaches means we can discover connections, correlations, and predictive biomarkers that were not apparent before. We are starting to use advanced algorithms or machine learning approaches to compile and combine heterogeneous data types—genetic, clinical, lifestyle, environmental—to build rich patient profiles and model individual-level outcomes with increased accuracy [6].

Thirdly, increased recognition of the importance of patient empowerment and shared decision-making in healthcare. Patients are no longer passive receptors of medical advice; they are active participants in their own care, seeking information, demanding personalized approaches, or even advocating for treatments that take their specific needs and preferences into consideration [7]. This paradigm shift towards more patient-centric care is highly congruent with the principles of precision medicine because, as the name suggests, at the core of this approach is the individual patient.

And last but not least is the pharmaceutical industry, which is also adapting to the precision medicine paradigm. The search for specific molecular targets and the design of therapies that are effective in genetically defined patient subgroups has given rise to the new field of drug development. Companion diagnostics are novel tests that help identify patients most likely to benefit from specialized drugs, and these are increasingly being used so that treatments are targeted and effective. Not only does this lead to more effective treatments for patients along the efficacy spectrum, it also reduces the chance that someone will experience adverse effects of drug treatment when they are unlikely to respond or predisposed to unwanted side effects.

At a high level, the shift from population-based, reactive models of health and healthcare to individual-based, proactive, and predictive care is one of the changes unfolding in healthcare. So, we are transitioning to a new era of healthcare made possible by advancements in technology, scientific discoveries, and growing societal expectations for more personalized and efficient healthcare. The era of personalized medicine and the resulting evolution of more individualized intervention and care stem from this excitement and the anticipation of a future where healthcare is tailored to the unique needs of each patient, one who is understood in the context of their own life, their behavioral and social environment, and their biology.

5.1.1 DEFINITION OF PRECISION MEDICINE

At its core, precision medicine is a new approach to disease prevention and treatment that takes into account individual variability in genes, environment, and lifestyle. The focus is no longer on disease therapy but patient therapy; the concept that no two patients are alike, and consequently, what is being recognized and targeted for best health outcomes needs to be personalized. Although precision medicine is sometimes confused with "personalized medicine," it is perhaps better termed "stratified medicine," since it classifies people into groups based on their distinctive features and then stratifies the intervention [8].

Individual-level data inform clinical decisions in the key tenet of precision medicine. This data can include things like genetic profiles (pharmacogenomics, genomics), molecular biomarkers (proteomics, metabolomics), imaging data, clinical history, lifestyle factors, and environmental exposures. Integrating and incorporating these different data streams together allow clinicians to better comprehend a patient's disease as well as predict how they may respond to a variety of treatments, enabling optimal treatment decisions for that individual patient.

Precision medicine is not only for treatment; it is also for prevention and early detection. Precision medicine predicts individual risk for specific diseases by correlating a person's genetic makeup or other risk factors to symptoms they will develop at a future stage. For example, those with a strong family history of breast cancer or BRCA gene mutations would benefit from enhanced surveillance, prophylactic surgery, and chemoprevention strategy. In the same vein, pharmacogenomic tests can detect individuals at high risk for adverse drug events, enabling the choice of alternative therapies or dose modification to prevent harm.

That we are heading towards precision medicine is a major paradigm shift in how we think about disease. Diseases have traditionally been categorized by the affected organ systems or clinical signs. On the other hand, in its description of disease, precision medicine acknowledges that diseases are not uniform homogenous disease processes; rather, there are multiple molecular subtypes of a disease that respond to different drugs in different ways. Cancer, once considered a singular disease, is now known to be a collection of hundreds of distinct diseases, each driven by different genetic mutations and molecular pathways. Precision oncology seeks to categorize cancers according to their molecular profiles and subsequently select targeted therapies that specifically target the fundamental drivers of each individual tumor.

Pharmacogenomics is important in precision medicine as it helps guide drug selection and dosing based on patient genetics. The response to a drug is significantly influenced by genes involved in drug metabolism, transport, and target pathways. The process of pharmacogenomic testing can help detect genetic variations that can indicate if a patient is likely to be a rapid, an intermediate, or a poor metabolizer of a drug, whether they are more or less likely to respond to a given therapy [9]. This information can also help clinicians to optimize it during drug therapy for maximal effectiveness in treatment and minimal risk of adverse drug reactions.

Gene editing technologies are also becoming powerful weapons in the precision medicine arsenal. Technologies ranging from CRISPR-Cas9 and TALENs to base editors enable the precise alteration of genes in living cells. Gene editing has incredible potential for correcting disease-causing mutations in genetic disorders, for developing novel cancer therapies targeting oncogenes, and for improving the power of immunotherapies by engineering immune cells as better tumor-targeting agents. The high specificity of gene editing allows for tailored therapeutic interventions.

Unlike the one-size-fits-all approach, precision medicine leans into heterogeneity and individual variation. Precision medicine not only goes far beyond the treatment of the average patient but also better categorizes the salient biological and clinical features of each patient to customize treatment for them. This paradigm shift necessitates an interdisciplinary workstream that incorporates aspects of genomics, bioinformatics, clinical informatics, and patient-oriented care. Precision medicine is not just a single technology or treatment modality but rather an all-encompassing approach to healthcare, making use of the latest data and technology to ensure patients receive the right treatment at the right time.

The potential of precision medicine alone is tremendous, including more precise diagnoses, better therapies, fewer adverse drug reactions, and anticipatory disease prevention. But it is only by overcoming incredible barriers such as ensuring that a robust data infrastructure is engaged, standardized protocols for data sharing are adhered to, that clinicians are educated regarding the use of such data, and the ethical/social implications are addressed that we can hope to realize this promise. Nonetheless, precision medicine is evolving in healthcare systems at breakneck speed, providing a new age of personalized, predictive, and participatory medicine [10].

5.2 GENE EDITING – REDEFINING THERAPEUTICS

Precision control over the genetic code of living organisms has been a goal for biologists for decades. Over the last few decades, however, this dream has gone from science fiction to just science, due to the revolutionary developments in gene editing techniques. Gene editing, or genome editing, is a breakthrough technology that has the power to revolutionize therapy, allowing for the treatment of the genetic underpinnings of diseases, rather than just their symptoms. This chapter section will provide an overview of this revolution at its heart, covering the technologies that can affect gene editing, the mechanisms by which they act, the challenging issue of delivery to target cells, and the current status of implementation of these potent technologies into the clinic [11].

5.2.1 GENE EDITING TECHNOLOGIES: UNLOCKING THE SECRETS OF THE GENOME

Gene editing itself is the capacity to make specific and well-documented changes to the DNA that is the blueprint of life. This process is accomplished by engineered nucleases, molecular scissors that can make incisions at designated points in the genome. The cell's innate DNA repair systems then go into action, and researchers can exploit these repair pathways to make targeted alterations at the location of the cut. Several promising gene editing technologies have evolved over the years, each with different pros and cons. Nonetheless, the CRISPR-Cas system is unsurprisingly well established and has taken over the field as a go-to tool owing to its simplicity, efficiency, and versatility [12].

5.2.1.1 CRISPR–CAS (CLUSTERED REGULARLY INTERSPACED SHORT PALINDROMIC REPEATS AND CRISPR-ASSOCIATED PROTEIN 9) SYSTEMS: GENOME SURGERY WITH PRECISE ACCURACY

1. **CRISPR-Cas9 (Short for Clustered Regularly Interspaced Short Palindromic Repeats and CRISPR-Associated Protein 9) is Widely Known as "Precision Genome Surgery:"** This system is derived from a genetic modification system of the immune system that is found naturally in bacteria and archaea to protect against viral infections. That system, a bacterial immune system, has been ingeniously engineered by scientists to serve as a powerful and versatile gene editing tool.
2. **CRISPR-Cas9 System Consists of Two Main Elements:** the Cas9 nuclease, which is an enzyme that acts like the molecular scissors, and a guide RNA (gRNA), a short RNA sequence that can be designed to correspond to the exact genomic locus that needs to be edited. The gRNA is complementary to a 20 bp sequence in the genome. It is the specificity of this sequence that enables targeted gene editing. The process starts with creating a gRNA that is complementary to the target DNA sequence. This gRNA is then complexed with the Cas9 protein. This complex is taken up into the cell, where gRNA leads the Cas9 nuclease to the correct site within the genome. When Cas9 and gRNA find their target, Cas9 introduces a double-stranded break (DSB) where the target DNA sequence is. This DSB occurs at a key point, activating the DNA repair machinery of the cell.
3. **Cells Generally Use Two Main Repair Pathways to Repair DSBs:** non-homologous end joining (NHEJ) and homology-directed repair (HDR). The NHEJ (nonhomologous end joining)

pathway is an error-prone but faster pathway that directly ligates the broken DNA ends. This process typically creates small insertions or deletions (indels) at the site of the cut, which can lose the sequence of the gene and effectively knock out its function. HDR, however, is a more accurate DSB repair pathway that uses a homologous DNA template to guide the final repair of the DSB [13]. To generate the necessary sequence changes, the CAS9 must be guided with an exogenous DNA template (containing the wanted sequence alteration) as well as the CRISPR-CAS9 system. Upon activation of HDR, the cell utilizes this template to mend the DSB, accurately inserting the genetic modification into the genome. This enables gene editing, gene addition, or other targeted changes.

4. **CRISPR-Cas9 is Simple and Elegant Game Changer in Gene Editing:** The process of designing gRNAs is fairly simple, and the system can be easily modified to target almost any gene in any organism. This versatility goes far beyond just knocking out or correcting a gene. Modified forms of Cas9 like base editors and prime editors have since been developed and represent even more precise and subtle genome editing capabilities that, that is, allow for single base pair changes without the creation of DSBs. Since its introduction, CRISPR-Cas9 has quickly overshadowed all previous methods to become the workhorse of the research lab worldwide while it also serves as the initial modality for the development of gene editing therapeutics [2].

5.3 PHARMACOGENOMICS: INDIVIDUALIZING DRUG THERAPY

This marks the beginning of a new era of Precision Medicine, which is all set to revolutionize drug therapy. For decades, drug development and prescription practices followed a "one-size-fits-all" approach, assuming that a typical patient with a specific condition will respond to a "standard" treatment regimen in a predictable way. However, what clinical practice and science continue to show is that this approach had its limitations. All humans will respond differently to medication; some are highly responsive while others feel nothing, or they might even have a serious adverse drug reaction. Pharmacogenomics, which investigates the contribution of genetic factors to variation in individual drug response, has arisen as an important field to address this variation and enable personalized drug therapy [14]. In this chapter, we discuss the genetic basis of drug response variability, the use of pharmacogenomic biomarkers in directing both drug selection and dosing, the implementation and clinical utility of pharmacogenomic testing, and describe examples of pharmacogenomic success stories from the management of a selection of diseases.

"Clinical reality for many years has been noting that people respond differently to the same drug, but why this is the case was unclear until the development of modern genetics and genomics science," said Marija Lavrnić, CEO of DigiKine. It is now well recognized that genetic factors underlie differences in the response to drugs and contribute to the variability in responsiveness and toxicity. This genetic basis for variability in drug response is derived from the inherent genetic diversity in the human population. Every person contains a unique array of polymorphisms which can affect drug disposition and action.

These genetic variations can influence drug response in several ways. First, genes encode proteins, making differences in genes translate to differences in protein structure, function, and abundance. For instance, many (likely most) of these proteins play a direct role in drug pathways such as drug-metabolizing enzymes, drug transporters, and drug targets (receptors [e.g., G protein-coupled receptors, nuclear receptors, etc.], enzymes, ion channels, etc.). Genetic variation in these genes can therefore directly modulate how a drug is processed by the body (pharmacokinetics) and how it exerts its therapeutic effect (pharmacodynamics).

Second, genetic variants can indirectly affect drug response through their effects on disease pathways. Disease itself has complex and heterogeneous genetic architecture. People with the same clinical diagnosis may have different genetic drivers of their disease. These genetic differences may show whether the disease responds to a given drug. Take cancer, for example, where tumors are characterized by a wide variety of genetic mutations. The particular mutations found in a given tumor can make it sensitive or resistant to targeted cancer drugs.

Next, genetic influences can interact with environmental or other lifestyle factors to alter drug response. The interaction of genes and environments are now acknowledged to be essential contributors to health outcomes, including drug response. For example, the impact of genetic variations in drug-metabolizing enzymes on drug metabolism can be enhanced in the presence of environmental factors that upregulate or inhibit these enzymes, like smoking or some food components [15]. Genetic makeup can also modulate the effects of lifestyle factors on drug metabolism and response, including diet, alcohol consumption, and physical activity.

Building on these early discoveries, pharmacogenomics is concerned with the discovery, characterization, and application of genetic polymorphism, which gives rise to variability in drug response. GWAS and candidate gene studies have enabled the identification of many pharmacogenes, genes that are associated with drug response. These studies have shown that hereditary factors, such as genetic variants in the genes encoding the drug-metabolizing enzymes, including the cytochrome P450 (CYP) enzymes, UDP-glucuronosyltransferases (UGTs), and thiopurine S-methyltransferase (TPMT), are significant determinants of pharmacokinetic variability. In the same manner, genetic variability in drug target encoding genes, including those of the receptors targeted by beta-adrenergic agonists, and the enzymes targeted by statins and ion channels inhibited by antiarrhythmic agents, have been noted to modulate pharmacodynamic variability as well [16].

It is important to go beyond the one-size-fits-all approach to drug therapy, and understanding the genetic basis of variability in drug response is critical for this process. Pharmacogenomic biomarkers, which are based on pharmacogenes and their variants, will enable the prediction of drug responses from an individual's genomic profile and guide the design of personalized therapeutic approaches [17].

5.3.1 PHARMACOGENOMIC BIOMARKERS IN DRUG SELECTION AND DOSAGE

Pharmacogenomic biomarkers refer to different genotypes that are predictive of patient responses to specific medications. Drugs may now be guided by biomarkers, which can aid in determining drug selection and dose, allowing clinicians to tailor pharmacotherapy to an individual's genome. Pharmacogenomic biomarkers can be classified into two major categories: pharmacokinetic biomarkers and pharmacodynamic biomarkers.

5.3.1.1 PHARMACOKINETIC GENES: EFFECTS ON DRUG METABOLISM AND DISPOSITION

Pharmacokinetic biomarkers are changes in genes that affect ADME (absorption, distribution, metabolism, and excretion) of drugs. These processes govern the level of a given drug at the target site that in turn impacts efficacy and toxicity. Polymorphisms in genes encoding drug-metabolizing enzymes and drug transporters are the best characterized pharmacokinetic biomarkers.

Drugs are biotransformed to metabolites (active/inactive), a process involving drug-metabolizing enzymes. CYP enzymes constitute a superfamily of enzymes that are central to the metabolism of the overwhelming majority of clinically used drugs. Information on very frequent polymorphisms in the CYP genes like CYP2D6, CYP2C19, and CYP2C9 which can change enzyme activity considerably. According to their CYP genotype, people can be categorized as poor, intermediate, normal (extensive), or ultrarapid metabolizers. Poor metabolizers, with low enzyme activity, have much higher drug levels with a risk of toxic (adverse) effects. Ultrarapid Metabolizers: Higher activity of the enzyme results in lower drug concentrations and may cause treatment to fail [18].

For instance, CYP2D6 is implicated in the biotransformation of many drugs, including antidepressants, antipsychotics, beta-blockers, and opioids, and so forth. Patients are classified as poor metabolizers of CYP2D6, which may cause side effects after dosing due to the drugs not being adequately broken down, and ultrarapid metabolizers, who may need higher doses to elicit a therapeutic effect. Pharmacogenomic testing of variants in CYP2D6 can be used to inform dose adjustments potentially leading to more effective treatment with fewer adverse events.

Membrane proteins known as drug transporters assist in the uptake and efflux of drugs through cell membranes, ultimately affecting the absorption, distribution, and elimination of drugs. Genetic polymorphisms in genes that code for drug transporters may influence drug concentrations in target tissues and systemic exposure; two such genes are SLCO1B1, which codes for the OATP1B1 transporter, and ABCB1, which codes for the P-glycoprotein transporter [19].

5.3.1.2 GENES INFLUENCING DRUG TARGETS AND PATHWAYS (PHARMACODYNAMIC GENES)

The pharmacodynamic biomarker is genetic variability in genes affecting target drugs and downstream signaling pathways. These biomarkers identify the probability of drug efficacy or resistance and can direct drug selection towards maximizing therapeutic benefit. Other relevant pharmacodynamic biomarkers are genotype differences in genes encoding drug receptors, drug-metabolizing enzymes, ion channels, and signaling molecules.

For instance, the target gene of warfarin, an anticoagulant drug, is VKORC1 (vitamin K epoxide reductase complex subunit 1). Genetic polymorphisms in VKORC1 and CYP2C9 (involved in the metabolism of warfarin) are important determinants of warfarin dose requirements. VKORC1 mutations influence the sensitivity of VKORC1 to warfarin inhibition, while CYP2C9 mutations regulate warfarin metabolism. Genetic testing for variants in VKORC1 and CYP2C9 is now standard practice to guide warfarin dosing, which greatly increases the safety and efficacy of warfarin therapy [20].

Pharmacodynamic biomarkers are crucial for guiding targeted therapy in cancer. Many cancer drugs are deliberately designed to take aim at specific molecular alterations, mutations, or gene amplifications that promote the growth of tumors. Testing for these target mutations through pharmacogenomics is critical to identify and select patients who will benefit from these targeted therapies. For example, EGFR tyrosine kinase inhibitors (TKIs) are beneficial for non-small cell lung cancer (NSCLC) patients with tumors that contain activating EGFR mutations. Screening for EGFR mutations has been incorporated into routine clinical practice in NSCLC to determine which patients are eligible for EGFR-TKIs. Likewise, therapies directed at HER2 are effective among breast cancer patients with amplification of the HER2 gene, and BRAF inhibitors are effective in melanoma patients with BRAF V600E mutations.

In addition to oncology, pharmacodynamic biomarkers play important roles in other therapeutic areas. By way of example, the genetic determinants of response to the beta-adrenergic agonists used in asthma is within ADRB2 (beta-2 adrenergic receptor). Polymorphisms in genes responsible for coding serotonin transporters and receptors can influence a response to selective serotonin reuptake inhibitors (SSRIs) used to treat depression. Pharmacogenomic testing for these pharmacodynamic biomarkers may help guide drug selection and enhance treatment outcomes for these conditions [21].

5.3.2 PHARMACOGENOMIC TESTING IN CLINICAL PRACTICE: IMPLEMENTATION AND UTILITY

Over the last decade, the clinical application of pharmacogenomic testing has seen continuous growth, driven by mounting evidence of its clinical utility and declining costs of genetic testing. Pharmacogenomic testing has been performed by various technologies, encompassing polymerase chain reaction (PCR)-based assays, DNA microarray, and next-generation sequencing (NGS). However, NGS-based approaches are becoming more common, synthetically characterizing multiple pharmacogenes at once, thus providing a comprehensive pharmacogenomic profile for a particular patient.

Pharmacogenomic testing may be adopted into clinical practice through various approaches, including preemptive testing and reactive testing. Preemptive pharmacogenomic testing consists of genotyping patients ahead of drug therapy as a routine component of clinical care, if possible [22]. The pharmacogenomic profile then resides in the patient's electronic health record and is accessible to inform the choice and dose of drugs in the event that a relevant medication is prescribed. Reactive pharmacogenomic testing occurs when a patient is already on a relevant medication and has an unexpected experience, such as loss of efficacy or an adverse drug reaction. Pharmacogenomic testing can subsequently be performed to assess whether genetic factors are playing a role in the observed response, and to direct treatment changes.

Pharmacogenomics testing has become clinically useful through studies and clinical practice guidelines. Pharmacogenomic testing enhances drug efficacy, decreases adverse drug reactions (ADR)/pharmacologic wariness, accelerates time to maximal therapeutic response, and decreases costs in the healthcare ecosystem across therapeutic areas. Pragmatic examples of how this could be used in psychiatry, for instance, pharmacogenomic testing can help select antidepressants and antipsychotics for patients, thus improving treatment success rates and reducing adverse effect risks. In cardiology, pharmacogenomic testing can optimize antiplatelet therapy following PCI and minimize the risk of stent thrombosis and bleeding events. Pharmacogenomic test application is critical for the personalized treatment in oncology to direct the use of targeted cancer therapies, improving patient treatment efficacy and survival outcome.

Nevertheless, even as the evidence supporting their clinical utility has grown, certain barriers to a broad implementation of pharmacogenomic testing in clinical practice remain. Challenges related to the implementation of pharmacogenomic testing in clinical practice include educating clinicians and training them in pharmacogenomics, integrating pharmacogenomic testing into clinical workflows and electronic health records, developing standardized reporting formats and clinical decision support tools, and addressing the ethical, legal, and social implications of pharmacogenomic testing. Pharmacogenomic tests offer valuable information to tailor treatments for individual patients, for which widespread insurance coverage is needed for equitable access to these tests, as reimbursement policies currently differ between health systems [23].

Addressing these implementation challenges will take a concerted effort from healthcare providers, researchers, policymakers, and payers. There need to be education and training programs designed to prepare clinicians to interpret and utilize pharmacogenomic test results in clinical practice. Pharmacogenomic testing can be integrated into electronic health records and clinical decision support systems, making pharmacogenomic testing results more readily available at the point of care. Standardized formats for reporting and clinical-utilization guidelines can enhance consistency and clarity in pharmacogenomic testing and interpretation. Ethical, legal, and social implications, including genetic privacy and potential discrimination, are important considerations for the implementation of pharmacogenomics. Advocacy for broader insurance coverage for pharmacogenomic testing is important to allow access to these tests for all patients who can be helped by personalized drug therapy [24].

5.3.3 SUCCESS STORIES OF PHARMACOGENOMICS IN MANAGING DISEASES

That's already translating into real clinical benefit in a number of different therapeutic areas with some of the strongest exemplars of success in disease management. Such success stories underscore the impact of personalizing drug therapy using genetic information.

A well-established and among the earliest examples is TPMT pharmacogenomics in decades of treating acute lymphoblastic leukemia (ALL) and inflammatory bowel disease (IBD). Thiopurine medications, including azathioprine (AZA) and 6-mercaptopurine (6-MP), are widely used to manage these disorders. TPMT is a drug metabolizing enzyme which inactivates thiopurine drugs. Patients with inherited TPMT deficiency have low TPMT activity, resulting in higher levels of active thiopurine metabolites and an increased risk of severe myelosuppression, a potentially life-threatening type of bone marrow toxicity. TPMT deficiency can be identified by pharmacogenomic testing for TPMT variants. Thiopurine-induced myelosuppression can be reduced significantly through dose optimization according to TPMT genotype, making thiopurine therapy more effective and tolerable. Information on TPMT, and often genes that are relevant to thiopurine therapy, is routinely recommended when initiating thiopurine therapy according to many clinical guidelines [25].

A well-known example is the pharmacogenomics of CYP2C19 in antiplatelet therapy with clopidogrel post-PCI. Clopidogrel is a prodrug that must be activated by CYP2C19 to have an antiplatelet effect. Variants causing loss-of-function for CYP2C19 are common (especially in East Asian populations), so individuals who have such variants express less CYP2C19 enzyme activity, leading to reduced activation of clopidogrel and thus a reduced antiplatelet effect. Patients carrying CYP2C19 loss-of-function variants undergoing PCI and managed with standard dosing of clopidogrel are at increased risk of adverse cardiovascular events, including stent thrombosis and myocardial infarction. Genotyping of CYP2C19 variants to identify patients who are clopidogrel poor metabolizers can identify patients upfront who need alternative antiplatelet therapy. Such patients might consider novel strategies for the prevention of thrombotic events, including higher doses of clopidogrel or traditional or alternative antiplatelet treatments (e.g., prasugrel or ticagrelor) to enable high antiplatelet efficacy with a reduced cardiovascular risk. Clinical practice guidelines for antiplatelet therapy following percutaneous coronary intervention (PCI) are becoming more increasingly integrated with CYP2C19 pharmacogenomic testing.

As such, in oncology, the application of EGFR pharmacogenomics in the treatment of NSCLC represents a prototypic example of targeted therapy that has been predicated by genetic biomarkers. Epidermal growth factor receptor (EGFR) tyrosine kinase inhibitors (EGFR-TKIs), including gefitinib, erlotinib, and afatinib, are efficacious in treating patients with NSCLC with tumors expressing activating mutations of EGFR, including deletions of exon 19 or L858R mutations. These mutations make EGFR sensitive to inhibition by EGFR-TKIs. For EGFR-TKI therapy in NSCLC, pharmacogenomic testing for EGFR mutations is required. EGFR-TKIs have shown superior progression-free survival and overall survival than chemotherapy in patients with EGFR-mutant NSCLC. EGFR pharmacogenomics has revolutionized the treatment paradigm for EGFR-mutant NSCLC, exemplifying the promise of precision oncology [26].

These and many other examples are what *in vitro* pharmacogenomics is clinically relevant in tailoring medications to improve the management of disease. With the development of pharmacogenes and their importance in the clinical aspect, the clinical application of pharmacogenomics is enhancing at an exponential rate with the wide availability of pharmacogenomic tests at an economical price. Genotype-guided drug prescribing will soon be a routine part of clinical care, leading to more accurate, effective, and safer drug treatment for all patients. Moving forward for a more personalized approach to medicine, pharmacogenomics will play a significant role in supporting this revolution in how we design, deliver, and dose medications so we can provide the right drug, for the right patient, at the right time.

5.4 MEETING REPORT: LEVERAGING GENE EDITING AND PHARMACOGENOMICS FOR SYNERGISTIC EFFECT IN MANAGING DISEASES

It is precisely the power of this synergy between individual technologies, as well as their independent contribution to a genetic risk factor, that makes precision medicine promise to be extraordinarily powerful. Together, gene editing and pharmacogenomics can provide a powerful strategy that allows the phenotypic effects of gene editing to be brought more fully into the therapeutic landscape, and the therapeutic landscape of gene editing provides a more accurate pharmacogenomic approach to genetics. The combined approaches maximally enhance therapeutic efficiencies, circumvent adverse events, bypass resistance, and tailor gene-based interventions by harnessing the synergistic twain of gene editing and pharmacogenomics. Combined with the innate field of gene editing and the patient-specific nature of pharmacogenomics, we are paving the way for a new frontier of precision in disease management.

5.4.1 COMBINED APPROACHES FOR GREATER THERAPEUTIC EFFICACY WITH REDUCED ADVERSE EVENTS

Combining gene editing with pharmacogenomics can provide a powerful strategy for simultaneously improving efficacy while minimizing the risk of adverse events. This is particularly challenging in traditional drug development and treatment paradigms where the natural variability in how patients respond leads to suboptimal efficacy in some individuals and intolerable toxicity in others [27]. Traveling with Gene Editing and Drug Response, one can solve both these problems.

The essential information required in understanding the relationship between genetic composition of human beings and drug response is supplemented by pharmacogenomics. Genetic variability can impact drug metabolism (pharmacokinetics) and drug targets (pharmacodynamics), thus resulting in a high degree of inter-individual variability in drug response and toxicity. Pharmacogenomic testing can identify patients who are likely poor responders to a specific drug or at higher risk of adverse drug reactions. This information enables clinicians to choose which drug and dosage is most appropriate for each patient, increasing the chance of clinical success and decreasing the chance of causing harm.

Broadly speaking, while current medicine relies on the use of drugs to indirectly target specific genes responsible for diseases, gene-editing technologies have enabled the direct manipulation of the genetic code itself, allowing for the correction of disease-causing mutations or even the enhancement of therapeutic responses at the genetic level. The therapeutic potential of gene editing is huge, but gene-editing application can be improved by integrating pharmacogenomic knowledge. Pharmacogenomic profiling can help guide the decision to use gene editing for a genetic disorder—for example, to provide technologically advanced practical drug metabolism capacity and sensitivity data of a potential patient. This knowledge can be utilized to custom-fit the pre- and post-gene editing therapeutic regimen so supportive pharmaceuticals can be selected and dosed appropriately based on a given person's genetic characteristics.

In such cases, particularly among complex diseases, the combination of gene editing and pharmacogenomics can prove beneficial, as the first targets the underlying genetic abnormality while the latter ameliorates the pathological state through pharmacological approaches. So, in the case of cystic fibrosis, gene editing intends to fix the root cause, the CFTR gene mutation, inside lung cells. Importantly, although gene editing can effectively halt the progression of a genetic disease, patients will still need pharmacological therapy for residual inflammation, infections, and other disease complications. Pharmacogenomics assists in selecting and determining the appropriate dosage of such adjuvant drugs based on each individual patient's genetic makeup, optimizing the treatment regimen while minimizing the chance of drug-drug interactions or harmful effects [28].

In addition, the pairing of the technology of gene editing and pharmacogenomics comprises a useful method for improving the effectiveness of classical drug therapies. To increase the sensitivity of patients to a certain drug, gene editing could be used to adjust the genes that affect their response to it. As an example in cancer therapy, gene editing could be employed to augment the expression of drug transporters in tumor cells, thereby increasing drug uptake and enhancing chemotherapy effectiveness. Alternatively, gene editing may allow silencing of genes conferring drug resistance to rescue drug sensitivity and circumvent treatment failure.

Enabling pharmacogenomic testing within the workflow of the gene editing therapy allows the clinician to better understand the biological milieu of the individual patient. This holistic approach in which gene editing is employed to correct the underlying genetic pathology of disease is complemented by pharmacogenomics informing clinical management of each patient before and after gene editing. These approaches create a synergy that results in maximized therapeutic efficacy, as the genetic defect is addressed by personalized drug therapy, and minimized adverse events, as drug selection and dosage is guided by an individual genetic profile. By combining the two approaches, the researchers took a step forward in precision medicine, offering a future where treatments are targeted not only at the disease but also at the individual patient.

5.4.2 GENE EDITING FOR ENHANCED PHARMACOGENOMIC RESPONSE

The emergence of drug resistance is one of the biggest challenges for the efficient treatment of many diseases like cancer, infectious diseases, and autoimmune diseases. There are different mechanisms that play a role in resistance, such as mutations in target genes, genes in drug metabolism pathways, and drug efflux mechanisms. One promising solution may lie in the emerging field of gene editing, which provides a novel and direct way to target and edit these resistance-conferring genetic changes. Integrating gene editing with pharmacogenomic strategy enables the design of precise action plans to reactivate drug sensitivity and increase treatment response in drug-resistant patients [29].

Pharmacogenomics is invaluable in candidate selection for the genetic basis of resistance to therapy. Pharmacogenomic testing can identify certain mutations or genetic variants linked to resistance against a specific drug. To illustrate with a cancer example, oncogenic or drug target genes can be screened for residue mutations associated with resistance by genomic sequencing of tumor samples. In the case of infectious diseases, analysis of the pathogen's genetics can identify mutations that confer resistance to antibiotics or antiviral drugs. Such pharmacogenomic information is critical to guide the use of gene editing to combat drug resistance [30].

Once a genetic mechanism of drug resistance is determined from pharmacogenomic testing, gene editing technologies can be used to directly target and correct these drug-resistance-conferring mutations. This can include restoring the original sequence of a gene in which a mutation resulted in drug resistance. For example, upon editing of a drug target gene that contains a mutation causing a reduction in binding affinity to the drug treatment, this mutation can be corrected through gene editing, and the original drug target sequence restored in cancer cells or pathogens, resensitizing them to the drug. CRISPR-Cas9 and similar gene editing technologies can be constructed to precisely target the mutated gene, and then they can be used to deliver an HDR template to populate the correct gene sequence, thus reversing the mechanism of resistance.

Gene editing to eliminate a drug resistance mechanism is a particularly promising approach in cancer therapy. At least half of cancer patients harbor mutations associated with resistance, showing that despite initial success, many targeted cancer therapies, which are only effective for a limited period, become ineffective through resistance mutations over time. Acquired resistance mutations can be targeted using gene editing, allowing sensitivity to be restored to the original targeted therapy or to other drugs that were inferior to the original targeted therapy due to acquired resistance. In EGFR-mutant non-small cell lung cancer, this can occur via secondary mutations in EGFR that confer resistance to EGFR-TKIs, which challenge the initial primary oncogene drivers. Genome editing strategies have been investigated for targeting these resistance mutations and resensitizing the tumors to EGFR-TKI, as well as prolonging the duration of therapeutic benefits.

One of the major threats to global health in infectious diseases is antibiotic and antiviral resistance. In the case of drug-resistant bacteria or viruses, gene editing can be employed to knock out and destroy those resistance genes, restoring the ability of the pathogenic organisms to be susceptible to antimicrobial agents. A particular technique, like CRISPR-Cas9, is used to target and remove specifically the gene responsible for antibiotic resistance in bacterial cells, thus restoring their sensitivity to antibiotics. Moreover, it is possible to use gene editing to specifically target and disrupt viral genes that produce drug resistance, thus overcoming antiviral drug resistance in viral infections.

5.4.3 PHARMACOGENOMICS TO INFORM GENE EDITING THERAPY: PERSONALIZING GENE-BASED INTERVENTIONS

Although gene editing provides a level of precision that is unparalleled in terms of genome modification, its therapeutic utility can be optimized through the context of pharmacogenomics. Pharmacogenomics is a field that has the potential to inform hot new approaches like gene editing therapy, where even small deviations from the genetic ideal would be magnified in each individual. Such tailored strategies can enhance the safety, efficacy, and long-term outcomes of gene editing therapies.

Data on individual variation in gene editing responsiveness not only improves on-target design but can also make predictions about individual responses to gene editing therapies themselves, paving the way for pharmacogenomic testing. Typically, gene editing works by modifying the DNA sequence itself; its approaches and methods are primarily focused at the level of the DNA. However, the cellular and physiological context in which gene editing occurs can also have an impact. Genetic polymorphisms in genes involved in DNA repair pathways, immune response, and cellular stress responses may impact the efficiency and safety of gene editing. Genetic predispositions of patients can also be influenced pharmacogenomically before gene editing therapy. These data could be used to customize individual gene editing protocols (e.g., dosage of gene editing reagents, adjustments to immunosuppression strategies) to maximize outcomes for individual patients.

In addition, pharmacogenomics can provide guidance on the most appropriate editing strategy to employ in individual patients. In some cases, different gene editing methods may also be available for treating a certain disease. For example, in the case of a genetic disorder brought on by a loss-of-function mutant gene, gene editing approaches could involve gene correction, gene insertion, or epigenetic editing to upregulate expression of the endogenous gene. Pharmacogenomic profiling can be used to guide the gene editing therapy option most likely to be effective and safe to use against the genetic background and the characteristics of the disease of an individual.

This technology can be used to optimize delivery methods for gene editing therapies in patients. As noted earlier, delivery systems play a key role in the success of gene editing. Pharmacokinetics of gene editing reagents, including viral vectors or nanoparticles, can vary based on genetic variations in drug transporters and drug-metabolizing enzymes. Patients with altered metabolism or transport of gene editing delivery vehicles can be identified through pharmacogenomic testing [31]; preferable delivery strategies can then be adjusted to improve gene editing efficiency and diminish off-targeting effects. If a subject is predicted to have rapid metabolic clearance of the viral vector, higher vector doses or alternative delivery methods can be considered to ensure adequate gene editing efficiency.

While gene editing acts at the genomic level, pharmacogenomics can help to inform not just gene editing itself but also the adjunct pharmacological therapies administered alongside gene editing. As already stated, gene editing is not necessarily curative in its own right, and patients will likely require a long-term pharmacological care plan for residual disease symptoms or complications. Pharmacogenomics may be employed to tailor these adjunctive therapies, ensuring selection and dosing are optimized based on the individual patient's genetic profile. Integrating pharmacogenomics personalized pharmacological management into a gene editing portfolio provides a broader spectrum of disease management strategies.

Simply put, pharmacogenomics is how we genomics is a lens into the world of the patient and the world of the drug. Integrating insights from pharmacogenomics in gene editing workflows opens opportunities to go from the one-size-fits-all model of gene-based intervention to personalized gene editing therapies. Personalization may also improve gene editing safety and efficacy, patient selection for gene editing studies, and ultimately, long-term therapeutic effect. With the clinical use of gene editing continuing to develop, the incorporation of pharmacogenomics will remain a cornerstone of fulfilling the promise of this genome editing technology in precision medicine.

5.5 CASE STUDIES: EXAMPLE INTEGRATED PRECISION MEDICINE STRATEGIES

Several case studies highlight the synergistic potential of combining gene editing and pharmacogenomics for disease treatment and are illustrative of current and future usage. These studies illustrate the use, or potential utility, of combinatorial treatments to improve therapeutic efficacy in multiple disease settings.

5.5.1 STATE FOR CASE 1: PERSONALIZED CHIMERIC ANTIGEN RECEPTOR (CAR) T-CELL THERAPY USING PHARMACOGENOMIC GUIDANCE FOR A PATIENT WITH LEUKEMIA

CAR T-cell therapy is an innovative type of immunotherapy in which the patient's T cells are genetically altered to recognize and eliminate cancer. Both hematological malignancies, especially leukemia and lymphoma, have been successfully treated with CAR T-cell therapy—however, treatment failure and severe cytokine release syndrome (CRS), a potentially life-threatening inflammatory response, can still occur. Pharmacogenomics applied to CAR T-cell therapy can offer more personalized treatment options for these patients, potentially improving treatment outcomes.

Measuring genomic variations for genes involved in leukocyte activation prior to CAR T-cell therapy can be used to identify high-risk patients for developing CRS, as well as patients at high-risk for low CAR T-cell expansion and/or persistence. One example includes an association of polymorphisms in genes of cytokines or immune response with higher susceptibility to the CRS. While not currently utilized in most settings, pharmacogenomic information can be employed to risk stratify patients for CRS and address preemptive strategies to prevent this complication, including adjustment of the conditioning chemotherapy regimen or prophylactic immunosuppressive agents.

In addition, pharmacogenomics can also be applied in the optimization of CAR T-cell design and production. Polymorphisms in genes associated with T-cell activation, proliferation, and cytotoxicity may also impact CAR T-cell effectiveness. Patient T cells can be profiled for pharmacogenomic genetic factors that might restrict CAR T-cell function. This is due to genetic limitations, which will then be addressed/overcome using gene editing technologies to improve CAR T-cell function. Gene editing, for example, may be exploited to improve the expression of co-stimulatory molecules in CAR T-cells or silence inhibitory receptors, improving both their anti-tumor effect and persistence [32].

Incorporating pharmacogenomics into CAR T-cell therapy can pave the way for a more patient-specific treatment approach, in which the risk of adverse events in patients is minimized while CAR T-cell response is maximized through genetic optimization followed by pharmacogenomically guided supportive care. Results suggest that this novel strategy could greatly enhance clinical responses to CAR T-cell therapy while also providing new avenues for approaching a broader variety of tumor types.

5.5.2 *CASE STUDY 2: GENE-CRACKING AND PHARMACOGENOMICS FOR INDIVIDUALIZED MANAGEMENT OF BETA-THALASSEMIA*

Beta-thalassemia is a hereditary hematologic disorder that is caused by a maximal number of mutations in the beta-globin gene, which ultimately heralds a diminished or deleted production of functional hemoglobin. Gene editing could potentially correct the underlying beta-globin gene mutations in patient hematopoietic stem cells (HSCs), thus offering a cure for beta-thalassemia. However, by integrating Pharmacogenomics, the gene editing approach for beta-thalassemia patients can be personalized!

In patients with β-type hemoglobinopathies, pharmacogenomic testing was leveraged to characterize the specific β-globin mutation present in each patient, which facilitates the design of gene editing reagents being precisely targeted to the mutation site. Depending on the beta-thalassemia mutation, these could include gene correction via HDR or base editing. It can also be used to identify genetic modifiers that affect the severity of manifestations of specific diseases and response to therapies. Genomic heterogeneity effects on beta-thalassemia phenotypes. Our findings clarify possible genetic determinants that can shape the clinical presentation of beta-thalassemia and response to gene editing. This information can provide insight on how best to dial in gene editing protocols and, more importantly, predict who will respond best.

In addition, pharmacogenomics can direct the pharmacological treatment of patients with beta-thalassemia receiving gene editing therapy. We also note that even post-genome editing, patients may require iron chelation therapy for iron overload, which is prevalent in beta-thalassemia patients [33]. The use of pharmacogenomic testing to identify genetic variations that affect iron metabolism and chelation drug response could facilitate personalized iron chelation strategies to maximize iron control while minimizing toxicity. Likewise, pharmacogenomics can help tailor the prescribing and dosing of other supportive medications, like erythropoiesis-stimulating agents or folic acid supplements, to the patient's genetic profile.

The gene editing and pharmacogenomics in the treatment of beta-thalassemia is a global precision medicine picture. Gene editing treats the underlying genetic cause of the disease, and pharmacogenomics ensures that both the gene editing and the pre- and post-relocation pharmacological management are personalized for each patient based on their specific genetic makeup and clinical characteristics. The twin approach has the potential to reach sustained disease freedom and enhance the quality of life in those affected with beta-thalassemia.

As mentioned previously, resistance to EGFR-TKIs poses a frequent issue in the management of EGFR-mutant NSCLC. Pharmacogenomic testing can identify the specific resistance mutations that develop in patients who progress on EGFR-TKI (tyrosine kinase inhibitor) therapy. Gene editing technologies can then be used to target these resistance mutations and restore sensitivity to EGFR-TKIs or other targeted therapies.

However, comprehensive pharmacogenomic profiling of tumor samples obtained from patients after the development of resistance to EGFR-TKI can identify the specific mechanisms of resistance, such as EGFR T790M mutation or MET amplification. Gene editing approaches can be developed to target these resistance mutations, as in correcting the EGFR T790M mutation or silencing MET amplification. This resistance could then be overcome, and the therapeutic effect restored by means of combination therapies of gene editing with EGFR-TKIs or MET inhibitors.

Besides, pharmacogenomics may develop to foresee which patients have a higher likelihood of developing specific resistance mutations and to inform strategy to prevent and/or delay resistance proactively [34]. Specific genetic variations in DNA repair genes or drug metabolism genes, for example, may affect resistance mutation risk. Pharmacogenomic information can help in stratifying patients by the probable level of resistance and guide empirical treatment selection to reduce the chance of developing treatment-rewarded resistant clones, such as, for example, in initial application of combination regimens or alternative orderings of therapies.

Gene editing and pharmacogenomics in the correction of EGFR-TKI resistance in lung cancer: a dynamic and adaptable approach to precision medicine. Pharmacogenomics directs how to discover resistance mechanisms, while gene editing provides tools to directly target and reverse these mechanisms. The process of pharmacogenomic tracking followed by genome editing interventions can allow for the continued efficacy of targeted therapies and beneficial management of EGFR-mutant lung cancer in the setting of acquired resistance [35].

The following case studies show the various and compelling manner in which gene editing and pharmacogenomics can be marshalled to improve management of disease. Integration of gene editing with pharmacogenomics allows us to create a personalized therapeutic approach, contextually focused at improving efficacy and minimizing unnecessary side effects, paving the way for a healthcare process of the future. Advancements in gene editing and pharmacogenomics are continuing, and together they represent a powerful combination of technologies that will change the future of medicine and usher in an era of precision healthcare.

5.6 CONCLUSION

Precision medicine was forecasted to lead us into a new era of healthcare. Now, it is clear that it is a real technology that has been unleashed. For more information about gene editing, pharmacogenomics, and how they're transforming the landscape of disease/care through their intersection, consider further research. With genome surgery and the personal nature of precision medicine through individual agent analysis, we find that the world is moving away from the past and finally back to a world that is healthier and more human, aware of the complexities of humanity and the things we need to overcome. Precision medicine is, at its core, a paradigm shift based on the interplay between advances in technology, basic science, and the realization that there is a natural uniqueness inherent in human biology. Genomics is moving away from the rigid standardization that has long dominated healthcare practice toward an understanding of the variables—constitutional genetics, environment, and so forth —that drive each individual's path forward with disease and its relationship to health. The four pillars of this new paradigm, known as proactive, predictive, personalized, and participatory (P4 medicine), can be summarized as follows: anticipate and prevent disease, predict individual risk and response to treatment, tailor interventions to individuals, and enable patients to become partners in their own care.

Gene editing and pharmacogenomics technologies are at the foundation of this emerging P4 medicine paradigm. Gene editing, which allows for unprecedented precision in altering the genetic code, has revolutionary potential to address causal genetic defects for inherited diseases, to target cancer-promoting mutations, and to develop new treatments for many conditions. From CRISPR-Cas9's "genome surgery" quality precision to a growing toolbox of base editors and prime editors, gene editing technologies are constantly advancing, becoming more efficient, more specific, and more versatile. They mark a paradigm shift from the symptomatic management of disease to target the underlying genetic etiology, with the potential for curative therapies for previously difficult-to-treat diseases.

In parallel, pharmacogenomics offers the personalized lens through which drug therapy can be optimized. Pharmacogenomics helps clinicians choose the right drug at the right dose for the right patient at the right time by understanding how an individual's genetic makeup affects his/her response to medications. Pharmacogenomic biomarkers, from pharmacokinetic genes that determine drug metabolism to pharmacodynamic genes that influence drug targets, are increasingly used to guide drug selection and dose adjustments in multiple therapeutic areas. Guiding everything from traditional thiopurine therapy approaches to safer and more effective therapy for leukemia, to the personalization of antiplatelet strategies in cardiology, to better guidance of targeted cancer therapies in oncology, pharmacogenomics is changing drug therapy from a trial-and-error-based approach to a genetically informed personalized strategy.

Yet the real magic of precision medicine is not in each of these technologies in isolation, but in their synergistic integration. Using a combined approach, integrating gene editing with gene-drug interactions through pharmacogenomics, is a powerful strategy that has the potential to maximize therapeutic efficacy and minimize adverse events. Through genomic editing, it is possible to modify the pharmacogenomic response and understand the mechanisms of drug resistance and drug sensitivity. On the other hand, pharmacogenomics is able to help gene editing therapy, customize precision of genetic intervention, and make it both safe and effective, based on a genetic profile of individuals. An exemplary approach of this synergistic interplay is adoption of personalized CAR T-cell therapy for leukemia, where a specialized pharmacogenomics-based approach can guide risk stratification, and gene editing can improve CAR T-cell functional profile (i.e., CAR T-cells used in this therapy) [1], whereas the treatment of beta-thalassemia realizes the approach where gene editing can ameliorate the genetic defect and pharmacogenomics can help optimize adjunctive pharmacological management. Gene editing and pharmacogenomics are being applied clinically in a wide array of disease categories. In perspective, precision cancer therapy is becoming a reality—pharmacogenomics determines therapy selection, while gene editing approaches are used to target cancer-driving mutations and mitigate drug resistance. For genetic disorders, gene editing promises to amend the underlying cause of disease, with pharmacogenomics leading the way to personalize and optimize gene-based therapies. In CV diseases, pharmacogenomics is customized for prevention and treatment strategies, and gene editing demonstrates new mechanisms to tackle genetic risk factors. Precision medicine strategies are implemented in infectious diseases to battle pathogen resistance; pharmacogenomics helps to personalize antimicrobial therapy; and gene editing provides novel approaches to directly target pathogens. Precision medicine is also opening new avenues in the inquiry of disease complexity and targeted intervention, even for complex conditions, which go beyond the monogenic disease. The journey to unlocking precision medicine's potential faces many hurdles. This is a data-oriented question, and ethical, social, and regulatory factors come first to mind; they should be conceptualized ex-ante. Navigating the ethical divide between somatic and germline editing will be no easy feat, and germline editing still faces intense ethical opposition as well as regulatory backlashes. The issue of equity and access is of paramount importance, emphasizing the need for deliberate actions to ensure equitable distribution of the potential benefits of precision medicine and to avoid worsening health disparities. The new era of genetic information demands alternative informed consent models that promote patient empowerment, autonomy, and privacy.

Only through carefully calibrated regulatory frameworks can gene editing and pharmacogenomics realize their potential to both facilitate life-saving clinical applications and allow for their use to proceed in safe ways for patients. Regulation differs worldwide, and international alignments are necessary to enable the responsible development and use of these technologies. The regulatory challenge is striking a balance between innovation and safety, creating flexible regulatory structures that can match the speed of scientific progress. When implementing precision medicine, societal implications, community perspectives, and cultural contexts should not be overlooked to meet public demands and ensure practical implementation to gain public trust and acceptance. Several aspects are necessary to transform clinical practice into integrating precision medicine into healthcare systems, which is paired with significant reengineering of healthcare delivery. Clinical workflows must evolve to integrate genomic testing and personalized therapies. Healthcare infrastructure will need to be improved to enable the analysis, storage, and sharing of data. Genetics, pharmacogenomics, and precision medicine principles must be integrated into the education of the healthcare workforce. The reimbursement models must be geared towards value-based care and encouraged to utilize precision medicine technologies. Keep up to date with predictive device ethics and regulations.

However, the efficacy of precision medicine to help achieve a healthier future is exciting and promising. It is a proactive vision, preventing illness before it occurs; predictive, foreseeing an individual's risk and response; personalized, crafting interventions to individuals' needs; and participatory, making patients partners in their healthcare. Such a future model for healthcare could lead to significant improvements in health, lower the cost of care, and better the quality of life for people and populations across the globe. Fulfillment of this promise will require concerted and united action by all stakeholders: researchers, clinicians, policymakers, patients, and the public. To advance gene editing and pharmacogenomics technologies clinically and overcome existing limitations, continued research and technology innovation is essential. Such collaboration across borders is essential for knowledge-sharing, alignment of regulatory standards, and addressing global health challenges around precision medicine. These technologies will foreseeably be used to contribute to, torch, provoke, or engage in acts of terror; create for some and destroy for others; cause harm or provide aid; liberate or oppress. The goal is to promote a policy development and health system transformation agenda to develop the infrastructure, workforce, and reimbursement mechanisms needed upfront to establish optimal infrastructure for precision medicine delivery through routine clinical practice.

Gene editing and pharmacogenomics will take precision medicine to an even greater level; without the ability to be effective, the technology, in general, would become useless. Unlocking this potential demands continuous efforts, collaboration, and a dedication to ethical principles and social responsibility. With precision medicine leading the charge, as we tackle the challenges awaiting us, a healthier tomorrow remains a tantalizing vision and inspiration propelling us forward on this incredible journey of medical and human advancement.

KEYWORDS

- **drug therapy**
- **gene editing**
- **genomics**
- **global health**
- **pharmacogenomics**
- **pharmacokinetic biomarkers**
- **precision medicine**
- **therapeutics**

REFERENCES

1. Relling, M. V., & Evans, W. E. (2015). Pharmacogenomics: Precision medicine and drug response. *Human Molecular Genetics, 24*(R1), R93–R99. https://pmc.ncbi.nlm.nih.gov/articles/PMC5682947/ (accessed on 5th January 2026).
2. Drozda, K., & Müller, D. J. (2024). Pharmacogenomics: Driving personalized medicine. *Expert Review of Molecular Diagnostics, 24*(6), 439–441. https://pmc.ncbi.nlm.nih.gov/articles/PMC10289244/ (accessed on 5th January 2026).
3. Mano, T. (2014). Pharmacogenomics in oncology care. *Journal of Clinical Oncology, 32*(5), 402–410. https://pmc.ncbi.nlm.nih.gov/articles/PMC3986526/ (accessed on 5th January 2026).
4. Juengst, E. T., McGraw, D., Henderson, G. E., & Green, R. C. (2016). Implementing precision medicine: The ethical challenges. *Genome Medicine, 8*, 14. https://pmc.ncbi.nlm.nih.gov/articles/PMC5217469/ (accessed on 5th January 2026).
5. Shane-McWhorter, L., & McWhorter, W. (2023). Personalizing personalized medicine: The confluence of pharmacogenomics, a person's medication experience and ethics. *Pharmacy, 11*(3), 101. https://www.mdpi.com/2226-4787/11/3/101 (accessed on 5th January 2026).
6. Kamenova, K., & Caulfield, T. (2023). Gene editing: The regulatory perspective. *BioChem, 3*(4), 96. https://www.mdpi.com/2673-8392/3/4/96 (accessed on 5th January 2026).
7. Baran, T. M., Foster, M. W., & McLeod, H. L. (2024). The future of pharmacogenomics: Integrating epigenetics, nutrigenomics, and beyond. *Genes, 15*(5), 877. https://pmc.ncbi.nlm.nih.gov/articles/PMC11677977/ (accessed on 5th January 2026).
8. Müller, D. J., & Zai, C. (2024). Clinical utility and implementation of pharmacogenomics for the personalisation of antipsychotic treatments. *Pharmaceutics, 16*(2), 244. https://www.mdpi.com/1999-4923/16/2/244 (accessed on 5th January 2026).
9. Nebert, D. W. (2019). Basic concepts in genetics and pharmacogenomics for pharmacists. *Pharmacy, 7*(4), 155. https://pmc.ncbi.nlm.nih.gov/articles/PMC6891005/ (accessed on 5th January 2026).
10. Cheng, E. H., & Sawyers, C. L. (2012). In cancer drug resistance, germline matters too. *Nature Medicine, 18*(4), 494–496.
11. Crew, K. D., Greenlee, H., Capodice, J., Raptis, G., Brafman, L., Fuentes, D., et al. (2007). Prevalence of joint symptoms in postmenopausal women taking aromatase inhibitors for early-stage breast cancer. *Journal of Clinical Oncology, 25*(27), 3877–3883.
12. Crews, K. R., Hicks, J. K., Pui, C. H., Relling, M. V., & Evans, W. E. (2012). Pharmacogenomics and individualized medicine: Translating science into practice. *Clinical Pharmacology & Therapeutics, 92*(4), 467–475.
13. Dreyer, N. A., Tunis, S. R., Berger, M., Ollendorf, D., Mattox, P., & Gliklich, R. (2010). Why observational studies should be among the tools used in comparative effectiveness research. *Health Affairs (Millwood), 29*(10), 1818–1825.
14. Jameson, J. L., & Longo, D. L. (2015). Precision medicine–Personalized, problematic, and promising. *The New England Journal of Medicine, 372*, 2229–2234.

15. Gillis, N., Patel, J., & Innocenti, F. (2014). Clinical implementation of germ line cancer pharmacogenetic variants during the next-generation sequencing era. *Clinical Pharmacology & Therapeutics, 95*(3), 269–280.
16. Gottesman, O., Kuivaniemi, H., Tromp, G., Faucett, W. A., Li, R., Manolio, T. A., et al. (2013). The Electronic Medical Records and Genomics (eMERGE) Network: Past, present, and future. *Genetics in Medicine, 15*(10), 761–771.
17. Henry, N. L., Giles, J. T., Ang, D., Mohan, M., Dadabhoy, D., Robarge, J., et al. (2008). Prospective characterization of musculoskeletal symptoms in early-stage breast cancer patients treated with aromatase inhibitors. *Breast Cancer Research and Treatment, 111*(2), 365–372.
18. Hertz, D. L., & McLeod, H. L. (2013). Use of pharmacogenetics for predicting cancer prognosis and treatment exposure, response and toxicity. *Journal of Human Genetics, 58*(6), 346–352.
19. Inglc, J. N., Schaid, D. J., Goss, P. E., Liu, M., Mushiroda, T., Chapman, J. A., et al. (2010). Genome-wide associations and functional genomic studies of musculoskeletal adverse events in women receiving aromatase inhibitors. *Journal of Clinical Oncology, 28*(31), 4674–4682.
20. Karapetis, C. S., Khambata-Ford, S., Jonker, D. J., O'Callaghan, C. J., Tu, D., Tebbutt, N. C., et al. (2008). K-ras mutations and benefit from cetuximab in advanced colorectal cancer. *The New England Journal of Medicine, 359*(17), 1757–1765.
21. Kelley, R. K., Wang, G., & Venook, A. P. (2011). Biomarker use in colorectal cancer therapy. *Journal of the National Comprehensive Cancer Network, 9*(12), 1293–1302.
22. Liu, M., Wang, L., Bongartz, T., Hawse, J. R., Markovic, S. N., Schaid, D. J., et al. (2012). Aromatase inhibitors, estrogens and musculoskeletal pain: Estrogen-dependent T-cell leukemia 1A (TCL1A) gene-mediated regulation of cytokine expression. *Breast Cancer Research, 14*(2), R41.
23. MacConaill, L. E., Van Hummelen, P., Meyerson, M., & Hahn, W. C. (2011). Clinical implementation of comprehensive strategies to characterize cancer genomes: Opportunities and challenges. *Cancer Discovery, 1*(4), 297–311.
24. Korngiebel, D. M., Thummel, K. E., & Burke, W. (2018). Implementing precision medicine: The ethical challenges. *Genome Medicine, 8,* 14. https://www.ncbi.nlm.nih.gov/pmc/articles/PMC5217469/ (accessed on 5th January 2026).
25. Hernandez, M., Cullell, N., Cendros, M., Serra-Llovich, A., & Arranz, M. J. (2024). Clinical utility and implementation of pharmacogenomics for the personalization of antipsychotic treatments. *Journal of Personalized Medicine, 16*(2), 244.
26. Venter, J. C., Adams, M. D., Myers, E. W., et al. (2001). The sequence of the human genome. *Science, 291*(5507), 1304–1351.
27. Shammas, M. A. (2011). Telomeres, lifestyle, cancer, and aging. *Current Opinion in Clinical Nutrition and Metabolic Care, 14*(1), 28–34.
28. Whirl-Carrillo, M., McDonagh, E. M., Hebert, J. M., et al. (2012). Pharmacogenomics knowledge for personalized medicine. *Clinical Pharmacology & Therapeutics, 92*(4), 414–417.
29. Johnson, J. A., Cavallari, L. H., & Touyz, R. M. (2013). Pharmacogenetics and cardiovascular disease—Implications for personalized medicine. *Pharmacological Reviews, 65*(3), 987–1009.

30. He, Y., Hoskins, J. M., & McLeod, H. L. (2011). Copy number variants in pharmacogenetic genes. *Trends in Molecular Medicine, 17*(5), 244–251.
31. Gaedigk, A., Ingelman-Sundberg, M., Miller, N. A., Leeder, J. S., Whirl-Carrillo, M., & Klein, T. E. (2018). The pharmacogene variation (PharmVar) consortium: Incorporation of the human cytochrome P450 (CYP) allele nomenclature database. *Clinical Pharmacology & Therapeutics, 103*(3), 399–401.
32. Zhou, S. F. (2009). Polymorphism of human cytochrome P450 2D6 and its clinical significance: Part I. *Clinical Pharmacokinetics, 48*(11), 689–723.
33. Beoris, M., Amos Wilson, J., Garces, J. A., & Lukowiak, A. A. (2016). CYP2D6 copy number distribution in the US population. *Pharmacogenetics and Genomics, 26*(2), 96–99.
34. Negrini, S., & Becquemont, L. (2017). HLA-associated drug hypersensitivity and the prediction of adverse drug reactions. *Pharmacogenomics, 18*(14), 1441–1457.
35. Reuter, J. A., Spacek, D., & Snyder, M. P. (2015). High-throughput sequencing technologies. *Molecular Cell, 58*(4), 586–597.

CHAPTER 6

Patenting the Genome: A Global Perspective of Intellectual Property Rights and Genetic Innovation

SOUFIANE OUARIACH,[1*] SAHIL LAL,[2*] BHUPINDER SINGH,[3*] and MANMEET KAUR ARORA[4*]

[1]*Abdelmalek Essaadi University, Tétouan, Morocco*

[2]*School of Law, Galgotias University, India*

[3]*Sharda School of Law, Sharda University, Greater Noida, India; University of South Wales, UK; Santo Tomas University Tunja, Colombia*

[4]*Sharda University, Greater Noida, Uttar Pradesh, India*

*Corresponding authors

ABSTRACT

The chapter provides a global perspective on this debated and convoluted field. It begins by placing the revolution in understanding genetics and the fledgling field of genetic innovation in the context of its enormous economic potential. The heart of the chapter unpacks the polarized arguments around gene patents. Arguments in favor of patenting are based on the notion of intellectual property rights as vital incentives for the generation, research, and development of biotechnology, as well as for attracting investments. In contrast, anti-patent arguments are addressed, including concerns over the monopolization of essential genetic data, restrictions imposed on healthcare and research, the possible emergence of a "tragedy of the anti-commons," and moral dilemmas in commodifying the human genome. Looking ahead, the chapter reflects on the potentially disruptive role of CRISPR and other forms of gene editing, the harmonization of

Genetic Editing for Healthcare: Advances in Disease Management.
Bhupinder Singh, Rishabha Malviya, Sahil Lal, and Sathvik Belagodu Sridhar (Eds.)

DOI: 10.1201/9781779643827-6

global patent law provisions, and emergent models of open science or data sharing as alternative solutions. It concludes with the notion that there is an urgent need for balanced policy frameworks that foster genetic innovation and equitable access, while also engaging ethical principles in this rapidly emerging field of biotechnology and genomic medicine.

6.1 INTRODUCTION

The decoding of the genetic code, the mapping of genomes in particular, the human genome, is one of the greatest scientific accomplishments of our age. This milestone has not only changed the way we view life itself, but has also created almost limitless possibilities for innovation in various sectors, especially in the field of healthcare. The applications of genetic data are revolutionizing medicine, from providing targeted therapies for conditions long deemed untreatable to generating advanced diagnostic technologies for the early detection of disease. But this transformative potential is mired in one of the most contentious and complex issues around: the patenting of genes and genetic technologies. Gene patenting, therefore, is a complex field influenced by scientific discoveries, economic incentives, ethical principles, and changing legal systems. In this chapter we introduce this complex space, which prepares the way for further elaboration on the global paradigm of intellectual property rights and genetic innovation [1].

6.1.1 FROM GENOME TO READ: THE CODE BEHIND GENES

The main target of the gene patenting debate, the gene, is the most basic unit of heredity. Genes, made of deoxyribonucleic acid (DNA), serve as the building blocks for life, providing the guidance needed for an organism's growth, function, and reproduction. For much of the 20th century, genes remained mostly abstract ideas, theoretical but unattainable at the molecular level. This paradigm shifted radically with the emergence of recombinant DNA technology, but more importantly, high-throughput DNA sequencing. This scientific adventure culminated with the Human Genome Project, an unprecedented international collaboration that ultimately resulted in the complete mapping of the human genome. This monumental undertaking, which was completed in the early 2000s, provided an unprecedented level of detail about the human genetic blueprint, identifying the estimated 20,000–25,000 genes that encode proteins and orchestrate cellular activities. Decoding the genome changed genes from abstract things to concrete information. Gene data, at its core, consists of the order of the bases of DNA, the difference within these sequences (polymorphisms) and the practical repercussions of the differences. This knowledge is not simply descriptive; it's deeply predictive, revealing an individual's risk for diseases, how they are likely to respond to drugs, and even their ancestry and physical characteristics. This has spawned a paradigm shift in biology and medicine by enabling us to read, interpret, and manipulate this genetic information. We can now make sense of our vulnerabilities to disease and tailor targeted strategies to prevent, treat, and cure the ailments that plague us in ways that were previously unimaginable. This transformation from viewing genes as biological units to perceiving them as an adjustable information system is at the heart of the discussion about gene patenting. For an invention to be patentable, it must be novel and non-obvious at the time of patent application, and so the question arises whether this information decrypted and comprehended represents simply a discovery in nature or constitutes an invention for which we could seek protection of intellectual property? This question is at the heart of the gene patenting debate [2].

6.1.2 GENETIC INNOVATION AND ITS ECONOMIC OPPORTUNITIES

The conclusions that scientists have reached after deciphering and manipulating genetic information have led to an unprecedented boom in genetic innovation. Discoveries and applications have mushroomed thanks to rapid developments in next-generation sequencing, gene editing tools like CRISPR-Cas9, and advanced bioinformatics analyses. In turn, advances in gene sequencing technologies have made it easier, cheaper, and faster to sequence an individual's genome, opening the door to large-scale genomic studies that have discovered genetic markers linked to a wide variety of diseases, from cancer and cardiovascular disease to neurodegenerative disorders and rare genetic conditions. This knowledge has led to the creation of genetic diagnostic tests that enable earlier and more accurate disease detection, personalized risk assessments, and targeted treatment strategies [3]. The most transformative field of genetic creativity may be gene therapy and gene editing. Gene therapy is a technique for correcting defective genes responsible for disease development. Among them, gene-editing technologies, and especially CRISPR-Cas9, provide us with unprecedented tools to make precise changes to the genome of living cells.

CRISPR, however, is an irreversible game changer with great potential that can work in treating genetic diseases and even for developing novel therapies in cancer immunotherapy by engineering cells for therapy. In addition, the area of personalized medicine, or precision medicine, relies heavily on genetic information. This means that drug responses can be predicted, and treatments or preventative strategies can be developed specifically for the genotype of the patient. These genetic innovations have huge economic potential. The biotechnology sector has emerged as a powerful economic engine worldwide, fueled by genetic R&D. Pharmaceutical companies are pouring money into gene therapies and genetically targeted drugs, seeing the opportunity for blockbuster treatments and cures. The diagnosis and treatment industry is thriving with genetic tests and cocktail programs being developed and sold for numerous health issues. Moreover, genetic technologies are being used beyond the confines of healthcare, for instance, in agriculture (genetically modified crops), industrial biotechnology (biofuels and biomaterials), and even consumer genomics (direct-to-consumer genetic testing). This economic potential is part of what drives the push for gene patenting. But according to companies and researchers, patent protection helps to recoup the massive investments required to conduct genetic research and development and also incentivizes further innovation in this fast-changing field [4].

6.1.3 INTELLECTUAL PROPERTY RIGHTS (IPRS): INCENTIVES FOR GENETIC RESEARCH

Intellectual property rights (IPRs), especially through patents, work by giving inventors exclusive rights to their inventions for a limited time, thus encouraging innovation. In the field of genetics, those who argue that genes and genetic technologies should be patentable do so with the claim that such monopolies are required in order to stimulate investment in this novel, high-risk, high-reward research. Creating a new genetic therapy or diagnostic test is a long, expensive, and often uncertain endeavor. Bringing a genetic innovation to market can take years of research and millions of dollars in investment, along with significant regulatory hurdles. Without patent protection, companies and researchers say, there would be little inducement to take on such risky and expensive projects. Competitors would be free to copy and commercialize their discoveries, choking off the financial basis of genetic research and perhaps curtailing future innovation [5].

Patents, in principle, offer a way to align private incentives with public good. Patents give inventors a limited monopoly on their inventions so that they can recoup their investments and profit from their inventions. The potential for profit incentivizes more research and development, and the continued advancement of genetic technologies will ultimately prove beneficial to society. They also encourage more inventions to be disclosed to the public. Inventors are required to publicly disclose their inventions in patent applications in exchange for patent protection. However, the inventors are of the view that public disclosure spreads knowledge and will lead to more innovation by allowing other scientists to improve on patented inventions or develop alternate methods. Just like an inventor may publish information about a new product, patent applications disclose data on things like rules or fragments of a gene, genetic markers, or novel methods; these data will reach the wider scientific community [6]. But there is also controversy surrounding the application of patent law to genes and genetic information. There are concerns that gene patents can stifle rather than encourage research, especially in the area of basic science. Patents on genes, key building blocks of life and products of nature, allow unfair monopolies over fundamental biological information, critics say. These monopolies, they argue, can limit the availability of genetic testing and therapies, drive up health care costs, and stifle future research by creating a "patent thicket" that slows innovation. The issue of gene patenting, therefore, is a manifestation of a deeper philosophical dichotomy: how to promote genetic innovation while ensuring equitable access to the fruits of genetic research and protecting the public interest.

6.2 FOUNDATIONAL CONCEPTS: GENETIC MATERIAL AND PATENTS

Gene patenting is a complex and long debate due to the very nature of patent law and also the unique nature of genetic material. To truly appreciate the nuances of this discussion, though, we have to build from early on the fundamental ideas of patent eligibility and where they meet genetics. In this section, we will explore these basic principles: what is a patent, the significant criteria of utility, novelty, and non-obviousness, and the key distinction between discoveries of nature and patentable inventions as especially relevant to patentable inventions that have been derived from DNA sequences, gene editing tools, and diagnostic methods. It will then describe the categories of patents that are applied to genetic material, thus preparing the reader for a thorough discussion of the international gene patenting scene [7].

6.2.1 PATENT: AN OVERVIEW

A patent is essentially a legal tool awarded by a government that allows an inventor exclusive rights to their invention for a limited timeframe, usually 20 years from the day the patent application is filed. This can be interpreted as the patent holder having the exclusive right to prevent others from making, using, selling, or importing the invention to the patent office. To receive this exclusive right, the inventor must publicly disclose their invention in enough detail such that others possess the relevant field knowledge to construct it. This quid pro quo aims to promote innovation by providing inventors with exclusive rights to their invention in return for sharing the knowledge with the public, and immediately after the expiration of the patent term, it enhances the public domain with new knowledge and technologies. In general, an invention must satisfy three basic requirements in order to be considered for patent protection: utility, novelty, and non-obviousness [8].

Second, it must be useful (utility) and have industrial applicability. It cannot simply be platonic speculation. Utility can be demonstrated if, in the context of a piece of existing genetic material, the sequence of a newly discovered gene has a function, such as coding for an expressed protein that may be a potential therapeutic or a diagnostic marker of a disease such as:

1. **Utility:** The utility requirement confirms that patents are granted on inventions that will have a real benefit to society rather than on discoveries without any apparent application.

2. **Novelty:** The invention has to be new as well as not known to the public or published in any way anywhere in the world before the date of filing of patent application. This is commonly known as "prior art." Establishing novelty for genetic material can also be difficult, especially with naturally occurring gene sequences. And though the gene itself occurs in nature, the patent application usually claims a particular form, like an "isolated and purified" piece of DNA. The question is whether this isolated form is "new enough" from the gene itself within the human body.

3. **Non-Obviousness (or Inventive Step):** An invention cannot be obvious to a person having ordinary skill in the art at the time the invention is made. This criterion is the most subjective and often litigated element of patentability. The patent application calls for patent examiners and courts to determine if the invention is a leap forward beyond the existing art and not simply a logical, predictable extension of what is already known. Non-obviousness in this case may focus on whether an expert in the genetic arts would have had little difficulty isolating the particular gene sequence or developing the particular genetic engineering technique taught in the patent at the time of the patent's application.

Forming the foundations of patent law are three basic criteria for invention: utility, novelty, and non-obviousness; all inventions, including those based on genetic material, are required to meet these test criteria. Instead, the application of these principles to genes and genetic technologies has been particularly difficult and controversial, leading to distinct challenges and interpretations in law.

6.2.2 WHAT ARE "INVENTIONS" IN GENETICS: PRODUCTS OF NATURE OR MAN-MADE?

One of the bedrock principles of patent law in most jurisdictions is the nonpatentability of "products of nature." Intuitively, this principle is based on the fact that natural laws, natural phenomena, and abstract ideas are the basis of science and technology development and should be in the public domain for everyone. Nevertheless, a clear distinction between a "product of nature" and a patentable "invention" is particularly nuanced and difficult to draw in the field of genetics. The human being has genes, and genes are indeed products of nature. They are naturally occurring sequences of DNA which evolved over millennia. Finders, after all, have long been considered basic discoveries of nature and patent-ineligible: just because you found a opened a new gene or its function, the scientific world has long considered all that a patent-less scientific achievement in itself. But patent law acknowledges that human intervention and genius can take a product of nature and elevate it into a patentable invention. Such transformation generally involves some type of human modification or change that produces an item that is "new" and "useful," and not found in nature in that unique form [9]. The key question in gene patenting then is the amount of human intervention that is necessary to convert a naturally occurring gene into a patentable invention. Patent offices and courts have struggled with this question in the past, leading to shifting interpretations and legal precedents. The early gene patents were typically directed to "isolated and purified" DNA sequences. The argument was that isolating a gene from its natural environment and purifying it and characterizing its sequence and function involved sufficient human ingenuity to make it patentable, even though the underlying genetic information itself was naturally occurring. It built on the notion that this isolated and purified form of the gene was not found in nature in that very form and had been given a novel utility through the agency of human beings.

But this "isolation and purification" rationale has been increasingly criticized and challenged in the courts, particularly in landmark cases like Association for Molecular Pathology v. Myriad Genetics in the United States. In Myriad, the Supreme Court ruled that isolated natural gene sequences are not per se patentable, even though they have been isolated. The Court reasoned that when a gene is isolated, it does not form anything new; rather, it simply reveals something that already exists in nature. The court explained that methods for isolating genes and (newly-discovered) applications of genetic information may be patentable, but the gene sequences themselves in their natural state are not patentable subject matter. The outcome of this and other legal rulings around the world have meant that creativity has turned to patenting more downstream inventions that involve these genetic materials, including new uses of genes, genetically modified organisms, as well as novel strategies in diagnostics or therapeutics. The difference now is that you have to show real human ingenuity—the human ingenuity needed to not just find a substance in the wild and separate it out. The invention must undergo enough transformation or manipulation of the natural product to merit any patent protection [10].

6.2.3 ALIENABILITY OF DNA SEQUENCES: ISOLATED GENES AND GENE EDITING TOOLS AS WELL AS DIAGNOSTIC METHODS

It becomes even more challenging to navigate the legal and ethical considerations surrounding gene patenting since a variety of genetic material and technologies may be claimed as patented, making it difficult to establish any clear guidelines for treatment. It is critical that, to help untangle this legal web, it is understood how different categories of genetic invention are treated to create patents like:

1. **Isolated Genes Complementary DNA (cDNA):** As previously described, many jurisdictions, notably after Myriad, have severely limited the patentability of isolated, naturally occurring DNA sequences. Although isolation and purification of genomic DNA is no longer sufficient for patentable subject matter, cDNA may still be patentable in various legal systems. cDNA is complementary to the mRNA, which only consists of the regions of the gene that encode protein (the exons), without the regions that do not code protein (the introns). The patentability of cDNA is often argued based on the understanding that it is not found in nature and must be produced at the hand of man. While this analysis indicates that cDNA falls outside of the exclusions from patentability, the actual breadth of cDNA patents will be determined based on continuing legal interpretation on the limits of patentability and may differ across jurisdictions.
2. **Gene Editing Tools (e.g., CRISPR-Cas9):** Technologies for gene editing, such as CRISPR-Cas9, are a new class of genetic inventions with both great promise and significant risk. Their tools enable precise and targeted change of DNA sequence within living cells. The patents on gene editing tools, on the other hand, are seen as being stronger than patents on isolated gene sequences. CRISPR-Cas9 and the like are undoubtedly the handiwork of humans; there is considerable ingenuity baked into their design and development. Patents on these tools usually describe the methodologies of gene editing, the engineered components of the system (e.g., Cas9 protein, guide RNAs), and specific applications of gene editing. But even with gene editing tools, patent disputes can emerge, especially with respect to the scope and breadth of patent claims and issues related to the naming of inventors when multiple research groups are involved in the development of an aspect of technology [11].

Methods of Diagnosis Utilising Genetic Information Gene patenting also extends to methods of diagnosis that are based on genetic information. The methods for doing so can vary from basic genetic tests that identify specific gene mutations to more extensive genomic analyses aimed at determining risk for disease or information on treatment decisions. The question of whether a diagnostic method claim is eligible for patenting generally turns on whether the claims are directed to a law of nature (e.g., a naturally occurring correlation between a gene mutation and a disease) or to a patent-eligible application of such a law. Just making a claim to a method of diagnosing something, in part, based on the observation of a natural correlation is likely not patent-eligible because it essentially wants to patent a natural phenomenon. However, if a diagnostic method entails the observance of a natural correlation but includes an inventive step, For example, a diagnostic method including common new methods of sample preparation, data analysis, or by therapeutic intervention based on the subject matter of the diagnosis, then the diagnostic method may be patentable. It is important to note that the patentability of DNA sequences and related inventions is subject to change and can be influenced by various factors, including legal precedents, technological advancements, and changing policy debates. Understanding this landscape is not easy, as it requires knowing the particular genetic material or technology being claimed, the type of patent claims at play, and the applicable law in the relevant jurisdictions.

6.2.4 TYPES OF GENETIC PATENTS: PRODUCT, PROCESS, AND USE PATENTS

There are patent types that are related to genetic make-up which can provide different protections and advantages as well as consequences in the fields of research, development, and commercialization. It is important to grasp these various forms to comprehend the extent and impact of gene patents. The most relevant patent types to genetic material are product patents, process patents, and use patents which described as:

1. **Product Patents:** These protect a physical object such as a gene sequence, a genetically modified organism, or a recombinant protein. In terms of gene patenting, product patents have traditionally been sought for "isolated and purified" DNA sequences. If granted, a product patent on a gene would give the patent holder exclusive rights to that particular DNA sequence, no matter how it is used. But, as noted above, the patentability of products having naturally associated gene sequences has been greatly curtailed. This seems to still hold true for those genetic entities that are truly novel, such as a synthetic gene or genetically modified organisms that are not found in nature, in which case product patents may still have value.

2. **Process Patents (or Method Patents):** These patents are for a specific way to do a given task. On the genetic level, process patents can extend to a variety of methods, including: DNA isolation and purification methods, gene editing methods, recombinant protein production methods, and diagnostic methods. Although process patents protect what sounds like the way in which a claimed method is performed, they do not inhibit others from performing the same steps by other processes to arrive at the same result. A process patent for a *CRISPR-Cas9* gene editing technique, for example, would bar its use only. Such a patent would, however, enable CRISPR specificity competitors to develop their own gene editing yet not be able to use that exact technique [12].

3. **Utility Patents (or Application Patents):** Use patents protect a new use of an existing product or process. In the field of genetics, it is common for use patents to cover the new therapeutic or diagnostic works based on a gene sequence or a genetic technology. For example, a use patent could issue on a known gene sequence if a researcher identifies a new, non-obvious therapeutic use for that gene. Use patents apply exceptionally when it comes to genetics as researchers are discovering more and more uses for already existing genes and genetic material to suit a wide range of different functions. Even if a use patent was granted to the discoverers, no one would be able to make that exact use of the patented gene sequence as described in the patent and perhaps not even that expeditious use of the patented gene sequence at all but anyone would still be able to use the same gene sequence for other, non-patented purposes. Different types of genetic patents offer varying levels of exclusivity and have different consequences for research and development. Broadly granted product patents can serve as a potential barrier to both entry and access to basic genomic information. As they are more narrowly defined, process and use-type patents can continue to incentivize innovation in specific methodologies and applications of genetic technologies. How the ideal balance between these different types of patent protection is achieved is an active topic of debate and policy discussion, as it seeks to balance the need for genetic innovation with the need to provide widespread access to the fruits of genetic research for the benefit of health care and the general population [13].

6.3 OVERVIEW ON GENE PATENTING: PRO, CON, AND AT A LOSS

The patenting of genes and genetic technologies has sparked what is arguably the most bitterly polarizing battle in all of intellectual property, with one side prattling on about the benefits of patent rights as the quintessence of innovation and the other condemning gene patents for stifling research, making healthcare unaffordable, and causing profound ethical dilemmas. This chapter section explores this controversy, unpacking the key arguments pro/con gene patenting. Recognizing these two "mindsets" will help us prepare for and deal with the changing world for ourselves and future generations with regard to the use of genetics. The debate is not just academic; it has real-world consequences for the pace of scientific discovery, the accessibility of life-saving diagnostics and therapies, and the very framework we use to think about and value the human genome.

6.3.1 THE PRO-PATENT THEORY: ENCOURAGING INNOVATION AND INVESTMENT

Most of the pro-patent option is arguing that a patent is an essential tool to enhance innovative ideas in genetic technology. The patent system exists to spur innovation, they argue, and without the incentive of the patent system, the immensity of financial risk and time associated with genetic research and development will dry up private sector investment and stagnate this critical frontier of scientific progress. They underscore the vital roles patents serve incentivizing research, garnering investment, and facilitating sharing of knowledge [14].

6.3.1.1 TO MAKE RESEARCH AND DEVELOPMENT IN GENETIC TECHNOLOGIES LUCRATIVE

Moving a genetic technology, be it a new gene therapy, an advanced diagnostic test, or a new gene editing tool, is inherently costly and time extensive. Research on genetics often requires years of basic scientific research, followed by pre-clinical study, then rigorous clinical trials, and navigating complex regulatory approval processes. The expenses at these stages can be mind-boggling, running into hundreds of millions or even billions of dollars. Furthermore, there is no surefire path to success. Many research projects never lead anywhere, and even if some promising discoveries might become commercially viable, they often don't do so [15]. In this high-risk, high-reward environment, patents are hailed as the necessary incentive for companies and researchers to pursue such ambitious and expensive endeavors. The argument is simple enough: Patents offer a period of market exclusivity, allowing inventors to recover their investments and profit from their innovations. The prospect of financial reward is viewed as a key driver for involvement of private sector players in genetic research and development. Without the financial incentive of patent protection, companies would be less inclined to do risky genetic research, worrying that rivals could swiftly imitate and profit from their discoveries without incurring the initial R and D expenses. Without adequate incentives, they argued, innovation would necessarily be stifled, and genetic technologies that could yield tremendous benefits for human health and welfare would slow to a crawl.

Take the case of gene therapies for rare genetic disorders. These therapies often focus on small patient populations and are therefore less commercially interesting than drugs for more common ailments. But the making of these medicines can be extraordinarily complicated and costly, requiring specialized know-how and advanced technology. In this context, patent protection can be especially important to encourage firms to commit to developing treatments for these under-researched and under-treated patient populations. While this exclusivity can be controversial, it is also what makes these niche markets commercially viable so that families of sufferers from rare genetic diseases are not left behind in the genetics revolution [16].

6.3.1.2 ATTRACTING PRIVATE INVESTMENT AND COMMERCIALIZATION OF GENETIC DISCOVERIES

Besides stimulating early research, patents are also essential to drawing in the private investment needed to convert basic genetic insights into commercially available products and services. Venture capital firms, pharmaceutical companies, and other investors have traditionally been risk-averse and want a clear path to a return on their investments. Patents grant this assurance by ensuring a protected market space for genetic innovations.

This development can often require scaling up production in the laboratory, doing clinical investigations, navigating regulatory challenges, and setting up marketing and distribution networks for a genetic discovery to move from laboratory bench to patient bedside. But these downstream activities can be even more costly than the initial research phase itself. In these advancing stages, private investors are more likely to part with capital if they believe the product or service will be insulated from competition and is likely to generate enough revenue to make their investment worthwhile. As a result, patents serve as an important intermediary between fundamental investigations into genetics and the consumption of new clinically useful genetic gadgets, allowing for the transmission of scientific progress to show insights for the rest of the community [17]. The biotechnology industry, a field that absolutely depends on genetic innovation, is proof of the power of patents to bring in private investment. One significant driver of the industry's growth and success has been the capacity to obtain patent protection for its genetic discoveries and technologies. That foundation created much of what is without question the largest industry boom the world has ever known, and companies like Genentech and Amgen and thousands of others have built off of strong patent portfolios to attract investment and develop genetic products with the goal of bringing them to market. Without the patent system, they argue, the biotechnology industry the world now knows and celebrates might not have come into being, and the rate of genetic innovation would have been slowed considerably [18].

6.3.1.3 THE UTILITY OF PATENT PUBLICATION TO ENCOURAGE DISCLOSURE AND KNOWLEDGE SHARING

Patents bestow benefits, yet they also play an important role in incentivizing the sharing of information and encouraging further innovation through their mandatory disclosure. Inventors must, in return for patent protection, adequately describe their inventions to the patent office, and so, an invention must be described in enough detail that someone skilled in that relevant field would be able to understand how to make or use the invention. This public disclosure is one of the fundamental aspects of the patent system and ensures that general knowledge does not remain undisclosed but is shared with the scientific community and society in general [19]. Once published, patent applications are a valuable source of technical information; if you have technologies of interest, you can monitor for relevant submissions. Such reviews provide in-depth insights about gene sequences, genetic engineering methods, diagnostic approaches, treatments that may be difficult to find elsewhere. This public disclosure also promotes scientific collaboration through avoiding redundant based research: scientists can take a look into patented inventions and learn sailing into uncharted waters that have not previously been powderingly sailed. In addition, they can stimulate new avenues of research and act as a basis for yet further innovations. This provides researchers with the opportunity to branch off from patented inventions, improve upon them, or simply discover new pathways, leading to faster developments and advancements in genetic science and technology. As such, the patent system is essentially a vehicle for sharing knowledge, even if it is within an umbrella of intellectual property. Patents are a form of temporary exclusivity but also are a contribution to the public domain because they disclose technical information. As it stands, the disclosure function of patents is particularly key in regard to complex and rapidly evolving fields, such as genetics, in which knowledge building or the sharing of knowledge to build upon previous discoveries are key to continued progress. In this perspective, the patent system is not only a private enterprise encouraged by government, but also a vibrant and creative ecosystem that promotes the long-term social good [20].

6.3.2 THE ANTI-PATENT COUNTERARGUMENT: WITHERING RESEARCH AND ACCESS

On the other hand, a strong counterargument to this line of reasoning says that patenting genes, especially if you consider gene sequences to be exactly the thing that was discovered, impedes research, limits access to potentially life-saving drugs and treatments, and raises even deeper ethical questions about the commodification of the human genome. Critics contend that gene patents grant monopolies over essential biological information, obstruct scientific advancement, and have an outsized negative effect on patient access to reasonably priced genetic testing and therapies [21].

6.3.2.1 MONOPOLIES AND BARRIERS TO ESSENTIAL GENETIC INFORMATION

Examples of gene patents: A major criticism of gene patenting is that they effectively grant huge monopolies on basic genetic information, with the result that portions of the human genome are "owned" by a private entity. Genes, critics say, are discoveries of nature, not inventions, and patenting them gives people exclusive control over naturally occurring substances and the information contained within them. They argue that this monopoly power can have harmful effects for research, healthcare, and public access to genetic information. If the number of tetraploid maize lines increases, it can greatly improve the proliferation rate of tetraploid maize lines. Gene patents can limit access to important diagnostic tests, as the patent owners may have the right to perform tests that are closely associated with the patented gene. Myriad Genetics' patents on the *BRCA1* and *BRCA2* genes, which are linked to breast and ovarian cancer, are one of the most well-known of these cases. Myriad's patents gave it exclusive control over BRCA testing in the United States, effectively barring other laboratories from offering competing tests even if they could do so at lower prices or using better methodologies. This was criticized as monopolistic and limiting access to affordable genetic testing by patients, and potentially restricting the advancement of other diagnostic methods. Nevertheless, the price of patented genetic tests is prohibitive, preventing access for many people and under-resourced health care systems, worsening health inequalities, and undermining preventative healthcare efforts.

In addition, gene patents can impede research due to a "reach-through" effect. Researchers studying diseases for which genes by patent are associated may have difficulty even if their studies are non-commercial or for developing new diagnostic or therapeutic approaches due to restrictions and licensing hurdles stemming from such IP regimes. Researchers may shy away from particular lines of research because of fear of patent infringement and because the process of negotiating the path through licensing agreements can be so difficult, especially in the gene patent arena. This chilling effect can hinder scientific discovery and stifle innovation in genetic technologies [22].

6.3.2.2 THE "TRAGEDY OF THE ANTI-COMMONS": PREVENTING DOWNSTREAM RESEARCH AND INNOVATION

The "tragedy of the anti-commons" describes a scenario in which many owners have the right to exclude others from a resource, resulting in underuse and inefficiencies. Critics warn that this troubling effect, observed in business, can be mirrored at the genetic level, where the proliferation of gene patents creates similar confusion in genetic research. An "anti-commons" effect appears to impede the exploration of research avenues that could otherwise yield significant outcomes. As the number of gene and genetic technology patents increases, researchers are increasingly forced to navigate this intellectual property thicket and license patents from multiple holders in order to conduct even the most basic research.

Passage through this "patent thicket" can be very cumbersome and expensive, as researchers have to dedicate significant time and resources to patent searches, licensing negotiations, and legal clearances. For academic researchers and smaller biotechnology companies, however, the transaction costs of acquiring numerous licenses can be daunting. This "anti-commons" effect can limit downstream research and innovation by making it hard and costly to access and use the genetic information and technologies needed to do so. It can also result in postponement in the development of novel genetic diagnostics and therapies, which ultimately slows down progress in the field and retards improvements to patient care. A wide covering gene patent, one which would cover not only specific gene sequences but also sequences, mutations, and applications related to those, is a particularly disastrous obstacle for research and innovation. Such wide-ranging patents can stand in the way of applications across whole areas of genetic research, restricting the ability of scientists to pursue new lines of investigation and explore new therapeutic options. The trade, "tragedy of the anti-commons" argument, suggests that gene patents can have the potential consequences of unnecessarily hindering the very innovation that they are supposed to foster, especially given their broad nature [23].

6.3.2.3 ETHICAL ISSUES OF THE COMMERCIALIZATION OF HUMAN GENOME

Those practical concerns about research and access lead to deep ethical objections related to the commodification of the human genome. People are the ones, not chemical compounds or industrial inventions, who get to decide that an entity is life, now or 20 years ago—they had better be right, because sitting on my lap, or your lap, is the life if we get another self, genetically identical or not; and genes are the way it is passed from generation to generation, and whether something is human or not, is the genetic information, it is the life, it is what defines us, is what makes the flesh unique. Patenting genes, they argue, turns vital sources of life into commodities that can be privately owned and commercially exploited. The encroachment of the human genome into the market raises profound ethical questions about the nature of human life and the worth of genetic material.

Opponents contend that genes are part of the common heritage of humanity and cannot be owned or controlled by private interests. To this way of thinking, patenting genes is like patenting organs or other basic components of the human body, which most people consider to be ethically unacceptable. The fear is that the patenting of genes will undermine the public perception of genes as common goods, which encourages the second perspective whereby the human genome becomes perceived as a set of proprietary components, bringing down the natural value of human existence and genetic variety [24]. Additionally, there are ethical concerns about the ability of gene patents to deepen already existing social and economic inequities. Patents prevent absorption of Essential Genetic tests and therapies? Will such responsibility be vested with evoking profit? Critics contend that the commodification of genes could prioritize commercial interests over the moral obligation of ensuring that all members of society have equitable access to the advancements arising from genetic research. The fact that we have a limited amount of data available to us at a given point (our data only goes up to October 2023) means that we are forced to work with what we have, and that the inherent methodology for how we interpret this data leaves room for individuals to interpret it in a manner that reflects their personal paradigm.

6.4 FUTURE DIRECTIONS AND CHALLENGES IN GLOBAL GENE PATENTING

The landscape of gene patenting is already complex and contentious, but it is likely to change further in the coming years. The future of intellectual property rights in this area will be determined by the quick pace of technological advancement in genetics, the evolution of legal interpretations, and the ongoing debate about ethical considerations. Approaching such a future requires new, proactive means of navigating the challenges this innovation presents, ensuring we maintain a balanced approach as the benefits of the tools become exaggerated against their growing inequities and ethical burdens. We will examine some key future directions and challenges facing global gene patenting, including the implications of new gene editing technologies, the push for harmonization versus the increasingly divergent nature of global patent law, the growth of open science and data sharing as alternatives to patenting, and the urgent need for policy recommendations to ensure a balanced and ethical gene patenting system.

6.4.1 *PATENT STRATEGIES FOR THE NEW GENETIC EDITING TECHNOLOGIES (CRISPR AND BEYOND)*

Since its discovery, the *CRISPR-Cas9* gene editing technology has revolutionized genetic research and has great potential in the field of therapy. CRISPR and other usable tool sets, like base editing and prime editing, have unparalleled speeds and range, allowing researchers to alter DNA sequences in highly explicit and targeted ways. These technologies are not only speeding up the pace of genetic research, they are also fundamentally changing the patenting landscape for genetic inventions. CRISPR technology is considerably simpler and cheaper than previous gene editing techniques, which has resulted in a boom of research and innovation in this field. That has created a messy and touchy patent landscape around CRISPR itself. The key patents covering CRISPR-Cas9 technology have been caught up in fierce and lengthy patent disputes, mainly between the University of California, Berkeley, and the Broad Institute of MIT and Harvard. These litigation wonder show the difficulties of establishing inventorship and the scope of patent claims in a rapidly evolving space where many research groups make essential contributions. The results of these patent disputes will have far-reaching consequences for the commercialization and availability of CRISPR products [25]. In addition to the initial CRISPR patents, the broad applicability of the effect gave rise to a slew of upstream invention downstream inventions covering novel CRISPR systems, more effective delivery, and specific therapeutic and diagnostic applications. Strategies for patenting in this space are becoming ever more nuanced, both for the core CRISPR technology itself and for a variety of experimental modifications, improvements, and uses that extend the level of expertise and inventions that came before. This spans patents on optimized Cas enzymes, new guide RNA designs, delivery systems, and CRISPR applications [26]. The arrival of CRISPR and other new gene editing technologies contradicts the traditional patenting paradigms. Innovation in this space is fast, which means patent applications must be filed and prosecuted quickly so that patents keep up with the technology. Moreover, the collaborative and open nature of much CRISPR research brings into play challenges to the applicability of traditional patent frameworks, predicated on individual inventorship and proprietary exclusivity. In the CRISPR era, the landscape of gene patenting is rapidly changing, with broader and more component-based patents being filed in the space, and there will be increased licenses to optimize improvements [27]. What it does require is a refresh on patentability standards and enforcement mechanisms to guarantee that the patent system spurs innovation instead of stifling research and access to these game-changing technologies.

6.4.2 HARMONIZATION VERSUS DIVERGENCE: GLOBAL PATENT LAW FOR GENETIC INVENTIONS IN THE FUTURE

Given that such research and development is global in nature, consideration has to be given to international patent law frameworks. Even though intellectual property rights are territorial (patents granted in one country are not automatically enforceable in another), there is an increasing acknowledgment of the need for increased harmonization of patent laws, particularly in the fields of biotechnology and genetics. Further harmonization will ultimately lead to a more predictable and uniform worldwide patent system, simplifying complexity and serving to reduce cost for innovators operating in multiple jurisdictions [28]. Several facets of international cooperation as well as regional agreements also exist, including the Patent Cooperation Treaty (PCT), which allows for the filing of a patent application internationally, and the Agreement on Trade-Related Aspects of Intellectual Property Rights (TRIPS Agreement), which sets out minimums for IP protection among World Trade Organization (WTO) member states [29]. Nonetheless, national and regional patent laws still differ radically, especially when it comes to gene patenting. Like what was mentioned beforehand, the US, Europe, and Asia have all taken a different approach to gene patentability based on their different legal traditions, ethical considerations, and economic priorities [30]. In the US, the Myriad decision severely limited the patentability of isolated naturally occurring gene sequences, whereas, although the European Patent Convention follows the same logic by excluding discoveries of nature, the rules of practice have been a little more flexible. While these regimes focus on the need to create a strong patent protection system, countries in Asia such as China and Japan face challenges in balancing incentivizing innovation with consideration for public health and affordable access to medicines. These divergences create challenges for companies seeking global patent protection for their genetic inventions, as companies are forced to navigate a patchworked mosaic of legal standards and patentability criteria.

There will be a tension between harmonization and non-harmonization, a fundamental unease that the future of patenting genetic inventions will hold closer to the vest of national and especially regional practices. Even as efforts continue to harmonize patent laws and their interpretations around the world, significant differences are likely to remain, reflecting the different values and frameworks prevailing in different jurisdictions. To the geneticist, this divergence in patentability requires a nuanced understanding of the idiosyncratic patent laws and practices of the major jurisdictions and a strategic approach to global patent portfolio management. Our shared understanding of the need for an internationally coherent patent system for genetic inventions will challenge such divergence, and dialogue among countries will be essential in order to navigate the needs of your population and the latest scientific advancements and ensure an internationally cohesive and effective system [31].

6.4.3 OPEN SCIENCE AND DATA SHARING: BUFFERING THE HURDLES OF PATENTING IN GENETIC RESEARCH

As calls for reform on gene patenting continue, alternative modes for stimulating and disseminating genetic research are emerging. Versions of open science, data sharing initiatives, and alternative forms of intellectual property are being explored as possible complements or even substitutes for traditional patenting in some areas of genetic research. Open science is a set of principles promoting transparency, accessibility, and collaboration in scientific research. For genetics, that includes sharing data, research materials, and research findings as broadly and as fast as possible. Open science is exemplified in large genomic databases, such as Human Genome Project data and databases maintained by the National Institutes of Health (NIH) and the European Bioinformatics Institute (EBI). Until now, researchers have been able to access huge amounts of genetic data in public databases and use it for their own research without worrying about patents and licenses [32]. Data sharing initiatives, which are promoted by funding agencies and research consortia, are becoming a standard practice in genetic research. These efforts promote or even require researchers to share their data, research materials, and software tools with the wider scientific community. This enhances reproducibility, speeds discovery, and enables collaboration, accelerating progress in the field. The open access movement in scientific publishing similarly follows open science principles, focusing on the right for free and unrestricted access to scientific publications, also in genetics. Attention is being paid to alternatives to traditional patenting, particularly for basic genetic discoveries and research tools. Some of these alternatives involve open licensing models, patent pools, or community-based intellectual property mechanisms. At the heart of this open licensing movement is the ability for patent holders to issue non-exclusive, often royalty-free licenses to their inventions, thereby facilitating broader access and use. Through patent pools, related patents related to a certain technology are aggregated, enabling collective licensing agreements and lowering transaction costs when navigating patent thickets. Instead of individualistic patenting approaches, community-based intellectual property models, based often in indigenous communities and developing countries, stress collective control and ownership with respect to genetic resources and traditional knowledge.

The move towards open science, data sharing, and alternative intellectual property systems is a response to the understanding that patents were never the best or most ethical way to promote innovation in every field of genetic research. Open access and collaborative models may be more amenable to accelerating scientific progress while also ensuring broad societal benefit for fundamental genetic discoveries and research tools. Patents still may be important for incentivizing the investment needed to develop and bring downstream genetic applications to market today, even for example, in the pharmaceutical and diagnostic sectors that require significant amounts of private sector research and development investment. Perhaps the future will be one of hybridization, with open science ideals applied to basic science, and smart patenting directed toward downstream commercialization but mindful of downstream impact, seeking balance between the models.

6.5 CONCLUSION

Despite quirks, the field of global gene patenting is even more complex, riddled with deep tensions, dynamic, and jettisoning into unknown territory. As we approach a genomic revolution, the governance of intellectual property rights of genetic material presents a key challenge. We have examined the complex arguments for and against gene patenting, reflected on the foundational legal concepts that inform this debate, and taken a world tour of different patenting laws and practices. Envisioning the future, one thing is apparent: the trajectory of gene patenting is likely to be influenced by the dynamic interaction of technological breakthroughs, legal developments, ethical discussions, and the pursuit of a system that strikes a balance between fostering innovation and ensuring access and welfare for all.

The fundamental tension that underlies the gene patenting debate is the conflict between the need to incentivize genetic innovation in the form of the promise of exclusivity in the market arising from a patent and the equally compelling need to compel equitable access to essential genetic information and technologies for all people. Advocates for gene patents are right that intellectual property rights are essential to funding risky, potentially high-reward genetic research. They argue that patents are necessary to recover the enormous expenses of research and development, to draw private investment for commercialization, and to stimulate the public dissemination of knowledge through patent disclosure. The reason for this was that without the prospect of getting protection through a patent on their findings, the argument goes, the rate at which innovative genetic work would flow would grind to a halt because vital diagnostics, therapies, and preventative measures were developed which have the potential to address some of the gravest challenges in human health. On the other hand, opponents of gene patenting counter with equally legitimate concerns about patents creating monopolistic control over basic biological knowledge, denying access to necessary healthcare services, and hindering downstream research and innovation. Monopolizing our genes, the very essence of being human, triggers ethical and societal fears of transforming our human genome into a commodity. These concerns, including limited access to genetic testing, the establishment of patent thickets through proliferating patents that would inhibit research, and the expected increase in health disparities, highlight the potential drawbacks of an overly broad or unregulated gene patenting regime. Such patents lead to the "tragedy of the anticommons," where the proliferation of intellectual property inhibits innovation rather than spurs it, and this potentially serves as a cautionary tale on the unintended implications that poor or overbroad patent protection in the field of genetics can have. The emergence of CRISPR-Cas9 and other groundbreaking gene editing systems has only made the patenting landscape more confusing. However, these tools have sparked complex patent disputes of their own and have raised novel challenges to patentability and enforcement. Its relative ease of use and low cost have democratized gene editing research, resulting in an innovation explosion, but also a more congested and contested patent space. These differing approaches to patenting, especially in the CRISPR age, are evolving to become very application-based, even down to specific improvements over previously patented techniques and may require a detailed understanding of the existing patent landscape and the often-arbitrary language associated with patents.

Internationally, attitudes on gene patenting are still varied, shaped by different legal traditions, ethical perspectives, and economic interests. International agreements and organizations are working towards a higher degree of harmonization, but there are still considerable differences between jurisdictions such as the United States, Europe, and Asia. In the US, for instance, the Myriad decision represented a dramatic departure from the patentability of isolated, naturally occurring gene sequences, and jurisdictions have applied their own nuances and interpretations. This can lead to global divergence, which means that for companies and researchers that operate in multiple countries, a nuanced understanding of specific national and regional patent laws is absolutely essential. Strategic patent portfolio management and a detailed understanding of the legal landscape in key market jurisdictions are essential to navigate this patchwork of regulations. Alternative models for promoting and disseminating genetic research are emerging in response to the challenges and limitations of traditional patenting. The emergence of open science projects, data-sharing platforms, and alternative models of intellectual property, such as open licensing and patent pools, may provide promising avenues through which to facilitate collaboration, accelerate discovery, and improve access to genetic information and technologies. These new policies recognize that, at least for primary genetic discoveries and research tools, open access and collaborative models may be more successful at maximizing social value than tight proprietary control. Nonetheless, patents likely still have a vital role to play in incentivizing the downstream commercialization of genetic applications, especially for industries such as pharmaceuticals and diagnostics, where meaningful private investment continues to be critical. A meritocratic hybrid—open science methods for upstream basic science and strategic patent strategy for downstream applications—might be the most pragmatic and effective way forward.

In summary, a sustainable gene patenting system will be achieved through a multi-dimensional approach. The need for this is of utmost importance when it comes to refining patentability criteria so that patents would only be granted for genuinely inventive applications and not on mere discoveries of nature. Data sharing and open science initiatives can facilitate research and reduce redundancy; mechanisms to provide access to patented technologies (e.g., compulsory licensing and patent pools) have been proposed as a way to address equity concerns. Both aspects are critical to ensuring that patenting practices around genes are aligned with societal values and ethical principles. These will definitely take some time, given the international cooperation gaps and harmonization challenges, but will eventually help deliver a relatively global patent system for genetic inventions that is more predictable and efficient.

In the end, the path forward for gene patenting will depend on our capacity to demonstrate wisdom, foresight, and a commitment to innovation and equity in addressing the complex and sometimes tenuous threads that make up this area of law and science all at once. The point is not to abandon any protection for patents but also not to give unconditional monopolies on genetic material. Instead, it is to create a more dynamic and tailored system that incentivizes responsible innovation, encourages the creation of life-saving technologies, and guarantees that the rewards of genetic research are available to everyone, regardless of their financial or geographic circumstances. It is through continued discussion, policy adjustment, and global cooperation on this issue that we will know if we are able to properly wield the transformative potential of genetic tools for the benefit of mankind, whilst avoiding their associated dangers and respecting the ethical concerns that must lead us as we enter the genomic era. Navigating this complex landscape demands ongoing reflection on our methods, an openness to adapt to emerging challenges, and a strong commitment to a future where genetics serve the greater good.

KEYWORDS

- **biotechnology**
- **gene patent**
- **genomics**
- **human genome**
- **intellectual property rights**
- **open science**
- **patent divergence**
- **polymorphisms**

REFERENCES

1. Simoncelli, T., & Park, S. S. (2015). Making the case against gene patents. *Perspectives on Science, 23*, 106–145. https://elsihub.org/collection/patenting-and-licensing-genomics-impact-access-genetic-tests-and-debates-about-patenting (accessed on 5th January 2026).
2. Park, S. S. (2014). Gene patents and the public interest: Litigating *Association for Molecular Pathology v. Myriad Genetics* and lessons moving forward. *North Carolina Journal of Law & Technology, 15*, 519–536. https://elsihub.org/collection/patenting-and-licensing-genomics-impact-access-genetic-tests-and-debates-about-patenting (accessed on 5th January 2026).
3. Park, S. S. (2018). The challenge to gene patents as feminist patent litigation. *Technology & Innovation, 19*, 659–670. https://elsihub.org/collection/patenting-and-licensing-genomics-impact-access-genetic-tests-and-debates-about-patenting (accessed on 5th January 2026).
4. N/A. (2023). *Patent Eligibility Restoration Act*, S. 2140, 118th Cong. https://elsihub.org/collection/patenting-and-licensing-genomics-impact-access-genetic-tests-and-debates-about-patenting (accessed on 5th January 2026).
5. N/A. (2024). *Patent Eligibility Restoration Act*, H.R. 9474, 118th Cong. https://elsihub.org/collection/patenting-and-licensing-genomics-impact-access-genetic-tests-and-debates-about-patenting (accessed on 5th January 2026).
6. N/A. (2013). *Association for Molecular Pathology v. Myriad Genetics*, 2013 WL 2631062. https://pmc.ncbi.nlm.nih.gov/articles/PMC3777541/ (accessed on 5th January 2026).
7. N/A. (n.d.). *35 U.S.C. §101*. https://pmc.ncbi.nlm.nih.gov/articles/PMC3777541/ (accessed on 5th January 2026).
8. Hall, J. M., Lee, M. K., Newman, B., Morrow, J. E., Anderson, L. A., Huey, B., & King, M. C. (1990). Linkage of early-onset familial breast cancer to chromosome 17q21. *Science, 250*, 1684–1689. https://doi.org/10.1126/science.2270482 (accessed on 5th January 2026).
9. Goldgar, D. E., Fields, P., Lewis, C. M., Tran, T. D., Cannon-Albright, L. A., Ward, J. H., Swensen, J., & Skolnick, M. H. (1994). Large kindred with 17q-linked breast and ovarian cancer: Genetic, phenotypic, and genealogical analysis. *Journal of the National Cancer Institute, 86*(3), 200–209. https://doi.org/10.1093/jnci/86.3.200 (accessed on 5th January 2026).
10. Wooster, R., Neuhausen, S. L., Mangion, J., Quirk, Y., Ford, D., Collins, N., et al. (1994). Localization of a breast cancer susceptibility gene, BRCA2, to chromosome 13q12–13. *Science, 265*, 2088–2090. https://doi.org/10.1126/science.8091231 (accessed on 5th January 2026).
11. Davies, K., & White, M. (1996). *Breakthrough: The Race to Find the Breast Cancer Gene*. John Wiley and Sons.
12. Wooster, R., Bignell, G., Lancaster, J., Swift, S., Seal, S., Mangion, J., Collins, N., Gregory, S., Gumbs, C., & Micklem, G. (1995). Identification of the breast cancer susceptibility gene BRCA2. *Nature, 378*(6559), 789–792. https://doi.org/10.1038/378789a0 (accessed on 5th January 2026).

13. Couch, F., Kamb, A., Rommens, J. M., Simard, J., Tavtigian, S. V., & Weber, B. L. (1995). *US Patent 5837492. Chromosome 13-Linked Breast Cancer Susceptibility Gene.*
14. Couch, F., Kamb, A., Rommens, J. M., Simard, J., Tavtigian, S. V., & Weber, B. L. (1995). *US Patent 6033857. Chromosome 13-Linked Breast Cancer Susceptibility Gene.*
15. Couch, F., Kamb, A., Rommens, J. M., Simard, J., Tavtigian, S. V., & Weber, B. L. (1995). *US Patent 6124104. Chromosome 13-Linked Breast Cancer Susceptibility Gene.*
16. Cook-Deegan, R. (2013). Are human genes patentable? *Annals of Internal Medicine.* https://doi.org/10.7326/0003-4819-159-4-201308200-00668 (accessed on 5th January 2026).
17. Kesselheim, A. S., & Mello, M. M. (2010). Gene patenting—Is the pendulum swinging back? *The New England Journal of Medicine, 362*(20), 1855–1858. https://doi.org/10.1056/NEJMp1004026 (accessed on 5th January 2026).
18. Walsh, T., Lee, M. K., Casadei, S., Thornton, A. M., Stray, S. M., Pennil, C., Nord, A. S., Mandell, J. B., Swisher, E. M., & King, M. C. (2010). Detection of inherited mutations for breast and ovarian cancer using genomic capture and massively parallel sequencing. *Proceedings of the National Academy of Sciences, 107*(28), 12629–12633. https://doi.org/10.1073/pnas.1007983107 (accessed on 5th January 2026).
19. Cho, M. K., Illangasekare, S., Weaver, M. A., Leonard, D. G., & Merz, J. F. (2003). Effects of patents and licenses on the provision of clinical genetic testing services. *The Journal of Molecular Diagnostics, 5*(1), 3–8. https://doi.org/10.1016/S1525-1578(10)60444-8 (accessed on 5th January 2026).
20. Merz, J. F., Kriss, A. G., Leonard, D. G., & Cho, M. K. (2002). Diagnostic testing fails the test. *Nature, 415*, 577–579. https://doi.org/10.1038/415577a (accessed on 5th January 2026).
21. Williams, H. L. (2013). Intellectual property rights and innovation: Evidence from the human genome. *Journal of Political Economy, 121*(1), 1–27. https://doi.org/10.1086/669706 (accessed on 5th January 2026).
22. N/A. (1980). *Diamond v. Chakrabarty*, 447 U.S. 303. https://pmc.ncbi.nlm.nih.gov/articles/PMC3777541/ (accessed on 5th January 2026).
23. N/A. (2010). *Association for Molecular Pathology v. U.S. Patent & Trademark Office*, 702 F. Supp. 2d 181. https://pmc.ncbi.nlm.nih.gov/articles/PMC3777541/ (accessed on 5th January 2026).
24. N/A. (2011). *Association for Molecular Pathology v. U.S. Patent & Trademark Office*, 653 F.3d 1329. https://pmc.ncbi.nlm.nih.gov/articles/PMC3777541/ (accessed on 5th January 2026).
25. Cook-Deegan, R. (2012). Law and science collide over human gene patents. *Science, 338*(6108), 745–747. https://doi.org/10.1126/science.1229854 (accessed on 5th January 2026).
26. N/A. (2012). *Mayo Collaborative Services v. Prometheus Laboratories, Inc.*, 566 U.S. _, 132 S. Ct. 1289. https://pmc.ncbi.nlm.nih.gov/articles/PMC3777541/ (accessed on 5th January 2026).

27. N/A. (2012). *Association for Molecular Pathology v. U.S. Patent & Trademark Office*, 689 F.3d 1303. https://pmc.ncbi.nlm.nih.gov/articles/PMC3777541/ (accessed on 5th January 2026).
28. N/A. (1948). *Funk Bros. Seed Co. v. Kalo Inoculant Co.*, 333 U.S. 127. https://pmc.ncbi.nlm.nih.gov/articles/PMC3777541/ (accessed on 5th January 2026).
29. N/A. (2010). *Bilski v. Kappos*, 561 U.S. ___, 130 S. Ct. 3218. https://pmc.ncbi.nlm.nih.gov/articles/PMC3777541/ (accessed on 5th January 2026).
30. Rai, A. K., Chien, C. V., & Clark, J. (2025). Molecular diagnostic patenting after *Mayo v. Prometheus*: An empirical analysis. *Journal of Empirical Legal Studies*. https://elsihub.org/collection/patenting-and-licensing-genomics-impact-access-genetic-tests-and-debates-about-patenting (accessed on 5th January 2026).
31. Duan, C., & Sarnoff, J. (2025). Certainly uncertain: An analysis of the Patent Eligibility Restoration Act of 2023. *American University Business Law Review*. https://elsihub.org/collection/patenting-and-licensing-genomics-impact-access-genetic-tests-and-debates-about-patenting (accessed on 5th January 2026).
32. N/A. (2025). *GENCODE 2025: Reference gene annotation for human and mouse. PubMed*. https://pubmed.ncbi.nlm.nih.gov/39565199/ (accessed on 5th January 2026).

CHAPTER 7

Biosafety Issues Relating to Genome Editing in Plants and Its Impact on Human Health

BHUPINDER SINGH[1,*] and CHRISTIAN KAUNERT[2]

[1]Sharda School of Law, Sharda University, Greater Noida, India; University of South Wales, UK; Santo Tomas University Tunja, Colombia

[2]Dublin City University, Dublin, Ireland; and University of South Wales, Pontypridd, Wales, United Kingdom

*Corresponding author

ABSTRACT

The development of plant genome editing technology, such as CRISPR-Cas9, creates opportunities to improve crop yield, nutritional value, and disease resistance. Nevertheless, it poses significant biosafety risks to both the environment and human life. Off-target effects are unintentional genetic mutations that could add new allergens or toxins to food crops, presenting potential health hazards. The chronic toxicology consequences of the consumption of genome-edited plants in the long term remain to be seen. There is a worry about the spread of genes from the edited plants to wild relatives, threatening ecosystems and biodiversity. The regulatory landscapes for evaluating genome-edited plants are different around the world, thus increasing the complexity of approval procedures. However, to ensure the safe use of this technology, it is necessary to carry out risk assessments and biosafety concerning food safety, environmental impact, and human health protection.

Genetic Editing for Healthcare: Advances in Disease Management.
Bhupinder Singh, Rishabha Malviya, Sahil Lal, and Sathvik Belagodu Sridhar (Eds.)

DOI: 10.1201/9781779643827-7

7.1 INTRODUCTION

Genome editing in plants has advanced the field of agricultural biotechnology and presents a novel arsenal for crop productivity enhancement, alteration of nutrition composition, as well as pest-, pathogen- or stress-resistance. The most popular method is CRISPR-Cas9, which revolutionized genome editing by being able to make specific changes in plant genomes. It holds the promise to tackle some of the biggest challenges facing food security around the world and feed a growing population as it enters this era. Nevertheless, the development of plants to be edited by genome technology has enormous potential impacts for human health/environmental security despite blooming benefits. This chapter examines these issues, assessing contaminants and different routes of introduction, considering the damage and the contemporary regulations for controlling genome editing unsafe use in plants [1].

The other crucial biosafety concern of GE is that edited genes might spread across wild relatives or non-target species; promiscuous hybridization can move the engineered gene into others. This gene flow may have ecological implications, such as the appearance of invasive species or atypical ecosystems. This course is with indirect effects on internal plant chemistry, impacting human health indirectly through food chains or environmental contamination. A single gene can have a large impact on an ecosystem and may be linked to multiple beneficial characteristics; therefore, the biosafety framework must take into account not only direct impacts of genome editing but spillover effects on biodiversity and ecosystem services as well.

On the regulation front, there are in a fractured global landscape. Although some countries heavily regulate genome edited plants as they do GMOs, for others, particularly those not containing foreign DNA, the level of regulatory scrutiny is relaxed. Various regulatory differences create obstacles to global trade, research harmony, and public acceptance of GM organisms. Therefore, this chapter highlights the urgent requirements for harmonized biosafety regulations that are scientifically accurate and flexible to be used while taking ethical as well as societal concerns into account in order to make sure genome-editing technologies developed responsibly and utilized optimally under agriculture. The innovative capacity of genome editing in agriculture and food safety is evident, but risk to biosafety concerns related to this technique will arise for its application in plant. To manage these predictable risks to human health/environment, enabling technologies must be proactively managed through appropriate regulatory mechanisms as well constant monitoring and mitigation measures [2]. This chapter would like to add its voice in that continued conversation by conducting a thorough examination of the biosafety concerns and regulatory dilemmas around GEPs, highlighting human health-related safeguards.

7.1.1 RELEVANCE OF THIS CHAPTER

The importance of biosafety in genome editing is a point-to-point issue because the release of genetically modified organisms (GMOs) or gene-edited plants can cause an unintentional response in the environment. Although plant varieties have been improved by traditional breeding for many years, genome editing changes are either more precise or remodeled faster at the molecular level. Such advances in technology continue to blur the neat boundary that previously seemed to distinguish conventional breeding from genetic engineering and, therefore, footage a reconsideration of existing biosafety protocols as well as regulatory frameworks thereunder. The major takeaway from work in this field is the risk of off-target effects potentially causing mutations that affect non-target (and often less well studied) regions within a plant's genetic make-up, with unknown consequences for development and fitness at an environmental level as much as human food safety [3].

Allergenicity and toxicity are significant human health concerns associated with genome-edited plants. New biochemical profiles could be created in plants by the use of gene-editing techniques, producing novel proteins or compounds that may elicit allergic reactions, toxic effects, and so forth, in humans. As an example, tweaking metabolic pathways to bolster the nutrient content of particular crops may also create toxic intermediates [4]. There are concerns about the potential impacts on human populations of gene-edited plants over a lifetime and numerous unknowns related to how these changes might impact our microbiomes, immune responses, or susceptibility to disease. Given the rapidly evolving genome editing technology, robust risk assessments and safety evaluations are required to minimize these risks.

7.1.2 GENOME EDITING IN PLANTS: OVERVIEW

Agricultural biotechnology has been revolutionized by genome editing in plants, which enables new precision to rewrite plant genomes and alter traits such as yield, pest resistance, or tolerance of environmental stresses. CRISPR-Cas9 (clustered regularly interspaced short palindromic repeats), TALEN (transcription activator-like effector nucleases), and ZFNs (zinc finger nucleases) are the most common genome-editing techniques used. These methods allow for changing a small number of base pairs in specific genes, providing very fine digital grain compared to conventional plant breeding. Because among all these methods, CRISPR-Cas9 is the best choice, as it combines easiness with efficiency and capacity of targeting multiple genes at a time.

It directs a Cas9 enzyme to an exact DNA location and cuts the double helix at these coordinates, allowing useful traits to be spliced in or harmful ones removed. TALENs and ZFNs, on the other hand, although old technologies, work in a similar way by cutting at defined DNA sites, producing breaks which are resolved via natural repair mechanisms of plants, resulting in genetic modifications. Where TALENs are modular, targeting is limited only by the available modules and not length, while ZFNs use engineered proteins to bind their required DNA. These include crops developed for increased disease resistance, enriched nutritional content, and improved adaptability to climate stress, all of which have considerable promise in terms of advancing sustainable agriculture as well as global food security [5].

7.1.3 SIGNIFICANCE OF GENOME EDITING IN AGRICULTURE

Genome editing has profound impacts in agriculture and offers innovative ways to customize our agricultural production systems, thereby helping us transact with global food security challenges, climate change issues, and so forth. With methods like CRISPR-Cas9, TALEN, and ZFNs, researchers can fine-tune the plant genome to create crops with enhanced characteristics, that is, increased yield, nutritional quality, as well as resistance towards pests/disease/abiotic stresses (entities in the environment that harm plants). Genome editing delivers faster, more precise advances than traditional breeding methods to develop crops that can thrive in poorly conducive soils and very wet conditions.

Genome editing can minimize the use of chemicals like pesticides and fertilizers, which incurs cost burden as well as environmentally strenuous approaches in agriculture. It improves food quality by increasing nutrient content, making crops safer to eat. It has the potential to transform modern agriculture, promoting sustainable development of the agricultural sector and food security for our growing world population by endowing farmers with resilient and high-yielding crops [6].

7.2 BIOSAFETY CONCERNS IN GENOME EDITING OF PLANTS

Together with other biosafety issues, the risks of these powerful new technologies need to be properly addressed for a safe application elsewhere. Genome editing technologies such as CRISPR-Cas9, TALENs, and ZFNs have come to the forefront when it comes to modifying plant genomes for numerous agricultural applications, from abiotic stress tolerance to disease resistance, but these new breeding techniques require extensive risk assessment due to their health hazards on humans and negative effects in the ecological environment that possibly endanger biodiversity. Although these technologies present promising solutions to global food security and agricultural issues, their escalating use poses major biosafety hazards, necessitating careful management.

The significant worry is the chance of off-target impacts. So, the idea behind genome editing is to make really accurate changes in the DNA molecule of a plant, but what happens when you edit that, as was learned before with immune response mechanisms? These off-target mutations have the potential to disrupt critical genes, which could result in unintended effects on plant development and performance, or even nutrient composition [7]. In practice, if the gene-edited crops themselves are ultimately consumed by humans and animals, downstream effects on human health may result from unintended proteins/compounds, which could be toxic or allergenic with chronic exposure. The concerns point to a requirement for thorough risk evaluations and testing procedures before pathogenetic crops are released into the environment or food chain.

Horizontal gene transfer, the possibility that edited genes could slip into wild relatives or other species by cross-pollination and natural processes, is another major concern. Greater than expected gene flow can result in significant unintended ecological consequences, such as the spread of invasive species or disruption of natural plant population genomic structure. If, for example, herbicide-resistant genes transferred to weeds and made "superweeds" that were more difficult to control, this could be bad because it would likely increase use of herbicides in the landscape, leading to unintended consequences on non-target plants or other types of harm. This case thus highlights the importance of monitoring gene-edited or driven plants and their ecological interactions in order to mitigate against this outcome [8].

Gene editing can change the metabolic pathways of plants so that they produce new proteins or chemical compounds, which may cause allergic reactions in humans and animals. The long-term safety of gene-edited crops still remains a large question mark, especially since genome editing can give rise to very subtle changes in the DNA code that may be missed by conventional testing methods. Consequently, gene-edited agricultural organisms for human consumption need to be reviewed just like any other GMO food products, and this requires comprehensive safety inspections of the plants created by altering their own genes [9]. Concerning that area brings up the environmental impacts. Crops that are gene-edited for pest or stress resistance may impact local biodiversity. For instance, crops engineered to express their own insect toxins can affect other insects that are not the target species (examples include pollinators such as bees and butterflies which have pivotal roles in ecological balance). Further, gene-edited crops released into the wild would have knock on effects for soil health and water usage in addition to disruption of other plant or animal species.

The regulation of genome-edited plants is also a big problem. Countries take different positions on how gene-edited crops should be regulated; some countries consider them similar to GMO, while others treat it more loosely, especially if no foreign DNA is added. This inconsistency poses several problems for international trade, consumer confidence, and public acceptance. This only adds a layer of complication to the work in developing an international framework for genome-editing technologies that boast proper and safe use on a global scale. It is note that while genome editing has a strong potential for improving agricultural productivity and sustainability, it at the same time presents significant biosafety issues warranting specific attention to prevent unintended risks either on humans and environment. Strict supervising, continuous research as well as monitoring systems is a pre-condition in order to perceive the advantages of genome-edited plants safely. A biosafety framework that effectively combines innovation with precaution will only be possible when scientists, regulators, and stakeholders work together [10].

7.2.1 UNINTENDED GENETIC CHANGES

Since genome editing in plants can result in off-target effects and unintended mutations, it may introduce genetic unsteadiness. Such unplanned effects could lead to negative consequences by interfering with critical functions of the plants, like growth, reproduction, and nutrient uptake.

7.2.2 ENVIRONMENTAL AND BIODIVERSITY ISSUES

Cross-pollination allows edited genes to infiltrate wild plant relatives and disrupt the ecology. Such gene flow could lower biodiversity, unsettle natural ecosystems, and upset the balance between plants. The biosafety impacts of these changes over the long-term with regard to the environment are becoming increasingly important.

7.2.3 HORIZONTAL GENE TRANSFER

Horizontal gene transfer is the movement of modified genes from plants to non-target organisms. Such gene flow would lead to the unintentional introduction of controversial traits into wild relatives that can be construed as a biosafety risk affecting natural gene pools and resulting in potential ecological disruptions.

7.2.4 THE DEVELOPMENT OF SUPERWEEDS AND PEST RESISTANCE

The spread of herbicide-resistant genes is an unintended consequence, whereas superweeds or pests that are resistant to management techniques could result. For example, agricultural challenges of weed and pest control may become more urgent in agriculture, needing enhanced intervention.

7.3 HUMAN HEALTH IMPACTS OF GENOME-EDITED PLANTS

Enter genome editing technologies, chief among them CRISPR-Cas9: these radical new tools are fundamentally changing the way that plant genomes can be modified. While these advances are exciting and certainly signify major steps forward in improving food security, nutrition, and disease resistance, since more genome-edited plants are getting to market, it is essential that we consider their impact on human health. Genome editing provides significant advantages; however, the unintended effects, allergenicity, and toxicity of new genes created via genome editing, and long-term health risks associated with it, need to be scientifically assessed and addressed. Availability of genetically edited plants is a major advantage, and it meets malnutrition associated with nutrient deficiencies, and so forth. Writers have engineered major food crops that now contain more vitamins, minerals, and essential nutrients than are endogenous to the unaltered plants. For instance, we can increase levels of beta-carotene using CRISPR technology in rice and corn, an effective way to counteract vitamin A deficiency, which presents a significant public health challenge faced by many developing countries [11]. Iron- and zinc-enhanced crops could also tackle global micronutrient deficiencies. Genome editing creates the opportunity to lower these cases by also achieving plants with improved nutritional profiles, genes for anemia, blindness, and impaired immune functions in terms of diseases linked to malnutrition.

Similarly, genome editing can also be harnessed to lower or do away with natural allergens and toxins in particular crops, thereby improving the safety of food. Some scientists have used CRISPR to modify peanuts and wheat, both of which can cause allergies in humans, so they could be harmless for people allergic to these foods. This technology offers a significant shift in the way we think about food-related allergic reactions and could potentially provide an end to this threat to the life quality for millions. In cassava, for example, reducing cyanogenic glycosides, as well as within potatoes to remove solanine, both of which are poisonous when consumed at high levels, genome editing can also lower harmful compounds. These changes could result in a more secure food supply, particularly in areas where these crops are mainstays of the daily diet.

Even though these potential advantages are exciting, it is necessary to continue assessing the unintended impacts of genome-edited plants on human health. Even though genome editing is very precise, sometimes it can still create off-target mutations, leading to unintended modifications of the plant's genetic structure [12]. These off-target effects may, in turn, result in the synthesis of completely new allergens or toxins that did not exist previously in this crop. Although the frequency of occurrence is much less compared to earlier genetic modification methods, modern genome editing techniques must still be carefully and closely monitored, in addition to thorough risk assessments, before we can consider them as safe for human consumption. Although there has not been any exhaustive health study to pay great attention to the hazards of eating genetically modified plants, few risks have been found when studied for only short periods. Critics suggest that short-term safety assessments may not reveal immediate harm, but the cumulative effects of using these plants for decades are unknown. There is also an ethical duty to regulate and label genome-edited crops. Transparent product labeling is fundamental to enable informed choice for the consumer, particularly when public uncertainty exists over long-term health effects. Genome-edited plants have enormous scientific potential in improving human health through enhanced nutrition and reduced allergens/toxins. Nonetheless, long-term research and strict regulation with public information are needed to ensure that any benefits these crops offer are not achieved at the expense of food safety or human health. The fate of genome-edited plants in our diet will come down to striking a careful balance between innovation and caution [13].

7.4 REGULATORY FRAMEWORKS FOR BIOSAFETY OF GENOME-EDITED PLANTS

Genome editing technologies: Almost two years ago, the unprecedented and rapid emergence of genome-editing tools such as CRISPR-Cas9 promised transformational advances in crop improvement for agriculture. These include boosting yield, enhancing nutritional value, and resistance traits for pests or disease via gene-edited plants that can also be created to tolerate droughts or temperature extremes, among other stresses. These advances have the potential to help resolve global food security and climate change dilemmas, but at risk are also likely biosafety threats [14]. The use of genome-edited plants in agriculture may pose significant environmental, health, and socio-economic risks, requiring adequate regulatory frameworks. These frameworks ensure that plants coming out of labs are safe for humans and biodiversity and played responsibly in global agriculture. Nonetheless, the regulatory environment for genome-edited crops is complicated and differs from country to region.

On a global level, the Cartagena Protocol under the Convention on Biological Diversity (CBD) is one of the most important regulations for proper usage of genetically modified organisms that include genome-edited plants. But even though these genome-edited organisms might be different from the GMOs of old, certain countries and international bodies treat them in much the same way. It stresses the precautionary measure, risk evaluation, and nation sovereignty in taking decisions at their own level, depending upon the threat to human health as well as the environment. It also provides transparency in the transboundary movement of GMOs, allowing importing countries to be informed and take decisions based on scientific risk assessments.

The guidelines of the Food and Agriculture Organization (FAO)/World Health Organization (WHO)'s Codex Alimentarius, developed by FAO in partnership with WHO (mentioned below), are intended to be used specifically as an elaboration on some aspects. In addition to these provisions concluded from national legislation, several international instruments serve as useful guidance: The Codex Alimentarius Guidelines for Safety Assessment of Foods derived from Genetically Engineered Plants provide recommended principles regarding food safety assessment. The guidelines hone in on looking into possible risks like allergenicity, toxicity, and unintended effects when genome-edited plants are to be introduced for human consumption as food [15].

7.4.1 NATIONAL AND REGIONAL REGULATION

1. **United States:** Regulation of genome-edited plants is comparatively lax. Genome-edited crops are largely regulated in the United States by the presence of foreign DNA, according to rules determined by The U.S. Department of Agriculture (USDA). As large numbers of genome-edited plants made using CRISPR and other processes may lack foreign DNA, they frequently fall outside the regulatory net set up for GMOs. In 2020, the USDA finalized a rule that exempts certain genome-edited crops from extensive pre-market evaluations if they could have been developed using conventional breeding methods. The FDA continues to maintain authority over food safety assessments, while the Environmental Protection Agency has jurisdiction for overseeing genome-edited plants that may be higher-risk in an environmental context—plants which confer pesticidal traits, from BT or CRISPR-Cas9-engineered pest resistant crops produced [16].
2. **European Union (EU):** The EU has been more circumspect. That year, the European Court of Justice declared genome-edited organisms to be subject to regulations every bit as strict as those for GMOs. As far as we can tell, this decision even pertains to crops developed using the kinds of techniques that do not introduce foreign DNA [17]. The precautionary principle is hardwired

into thinking in the EU, and so extensive risk assessments precede even labeling in order to guarantee full transparency for the consumer. But critics maintain that the directive might dampen innovation and impede the potential genome-editing benefits for European agriculture—improved food security, climate resilience.

3. **Canada:** The Canadian Food Inspection Agency (CFIA) and Health Canada regulate novel plants based on the end product, not whether or how new technologies are used in their breeding. With this, genome-edited crops can be returned to a case-by-case consideration for risk assessment of biotech traits instead of the current one-size-fits-all GMO regulatory oversight. A genome-edited plant that does not express new traits beyond what is already present will likely require fewer regulatory approvals [18].

4. **Japan:** Differentiating genome-edited plants that do and do not contain foreign DNA, Japan simply opted to base their regulation on the presence or absence of added genes. When it comes to plants that do not contain foreign DNA (as is commonly the case with CRISPR-based modifications), they are not nearly as regulated as GMOs. While genome-edited plants must still be reviewed for environmental and food safety to determine whether they might present hazards to human health or the environment.

7.4.2 PUBLIC CONSULTATION AND RISK ASSESSMENT

Risk assessment is an indispensable part of the regulatory frameworks that cover genome-edited plants. Even in countries which allow commercialization, high-impact plants will need to undergo a science-based process for comprehensive assessment of health and environmental safety effects before the plant can be offered as food/feed or used for processing [19]. These may include evaluations of allergenicity, toxicity, unintended effects, and ecological consequences such as gene flow, leafy relatives, or non-target species. Post-market surveillance is also carried out by regulatory agencies to track long-term effects. Biosafety regulation also increasingly takes into consideration the value of public engagement, along with traditional scientific risk assessments. The arrival of genome-edited crops has resulted in ethical, cultural, and socio-economic issues such as food sovereignty and biodiversity protection. Thus, it is for this reason that most regulatory frameworks promote or mandate the need for public consultations and stakeholder input to discuss the benefits and risks of genome-edited plants in the open. Public trust will be essential for the successful adoption of these technologies, which means a transparent regulatory process is itself an important piece [20].

7.4.3 FUTURE PERSPECTIVES AND CHALLENGES

The regulatory landscapes of genome-edited crops are still developing; however, considerable practical challenges persist. A key one is the absence of a coordinated or harmonized global regulatory regime for genome editing, which can in turn create trade barriers and restrict acceptance of edited crops. Nations will also have to consider ethical concerns, feeding practices, and the growing consumer call for clear-cut labels [21]. As the science and technology continue to progress, regulatory frameworks will need to strike a balance between desirable innovation that could allow for realization of benefits from genome-edited plants while also minimizing any unnecessary (and unproven) risks associated with health or environmental factors. Given the multiplicity of cultures and economies dispensing food across global markets, cooperation on an international basis in setting standards will be essential for establishing both responsible use as well as broad acceptability around what should constitute a genome-edited crop [22].

7.5 ETHICAL AND SOCIAL CONSIDERATIONS IN BIOSAFETY AND GENOME EDITING

The successful application of genome editing technology (with its more sophisticated techniques such as CRISPR-Cas9) to agriculture and biosafety has raised significant ethical and social implications [23]. The technology has many realistic values for food security, environmental sustainability, and human health improvement but also brings some complex challenges with the need for certain ethical considerations that should be tackled cautiously from a societal perspective. They could mean the end of hunger and malnutrition if they produce higher crop yields for food security or improve the nutritive value of staple foods, including grains such as rice and soybeans. However, there are fears that genome editing will only serve big agricultural corporations in richer countries with industrial-scale agriculture systems. There would, however, be a challenge faced by small-scale farmers, particularly in developing countries, of not being able to access or afford these innovations. That could even deepen the difference between industrialized and smallholder agriculture, which would create social inequities [24]. Leveraging genome editing to underserved and low-resource communities with devising ways that the benefits of genome-editing are equally distributed across all sectors is a paramount ethical obligation.

A second worry is the environmental consequences of editing genomes in organisms. Breeders could accidentally affect ecosystems by breeding new traits into crops, causing the quality of natural habitats to change or threaten wild plant biodiversity because the edited genes would spread themselves far and wide. This, in turn, raises questions over the ethical duty that human society may have to protect natural ecosystems and biodiversity. Also, just as rational design can result in crops resistant to diseases and pests, potentially the real driver for farmers, genome editing runs a risk of generating what could be called "super nuisance" or other organisms that will become resistant.

Transparency should inform consent. Indeed, consumers have the right to know that they are eating genome-edited food so as to be able to choose what foods they can eat carefully. This has sparked controversies about whether genome-edited foods will be labeled, some arguing that labeling should allow consumers to choose and maintain confidence, while others think labeling may suggest a stigma against beneficial developments unnecessarily. Public engagement and transparent communication are key components for alleviating societal anxieties, instilling trust in genome-edited products [25]. Genome editing also clashes with cultural and religious beliefs. There are also moral objections to changing the genetic structure of plants and animals that result from interfering with nature or going as far as some religious teachings allow when it comes to playing God. It is bread and butter that socio-cultural factors must better inform the development of genome editing technology and its application in ways that do not overshadow diverse cultural outlooks. Altogether, ethics and societal aspects of biosafety genome editing are 3-dimensional. Integration of innovation with caution, equity, and respect for diverse values are central to the responsible use of genome editing.

7.6 CASE STUDIES ON GENOME-EDITED PLANTS AND HUMAN HEALTH

Genome editing, especially the CRISPR-Cas9 system-based technology, has revolutionized plant breeding via precise genetic modification. This includes having crops with better nutritional content, as well as improved resistance to pests and diseases or tolerance of adverse environmental factors. Despite the significant advantages these trees offer in agriculture, their influence on human well-being is still an area that requires further exploration. In this section, the current human health risks that arise from dual editors of genome-edited plants are proposed, and pathways contributing to genotoxic effects by dual-editor-mediated off-target cleavage events formation after consumption [26].

7.6.1 CASE STUDY 1: VITAMIN A DEFICIENCY AND GOLDEN RICE

The best-known genetically modified (GM) crops, golden rice, has been improved with genome editing. It was formulated to combat vitamin A shortage, which is a widespread health issue affecting rice-consuming countries in the developing world. Vitamin A deficiency results in blindness, reduced immune function, and increased mortality associated primarily with young children and pregnant women. Golden Rice was biofortified to produce beta-carotene, an element that the body can convert into vitamin A using our in-house kitchen machinery. Beta-carotene levels in early GM versions of Golden Rice were low, and resistance from some end users towards GMOs was a major issue [27]. Now, with a little nudging from genome editing, the beta-carotene will be abundant to serve as an effective weapon against vitamin A deficiency. Golden Rice itself, when eaten in 150 g serving size (about the wallop of a tennis ball), offers about 60% of daily vitamin A for children as well.

Golden Rice delivers an important public health outcome for human nutrition. Still, questions linger over the real effects of genetic or genome-edited food consumption on people decades from now. Golden Rice has gone through extensive safety assessments, and all have found no health risks linked to this improved variety; however, controversy over the potential lack of human safety on GMO crops slows down its adoption. In the debate over Golden Rice and other genome-edited crops, ethics still play a key role, as consumers have every right to make an informed choice about their food [28].

7.6.2 CASE STUDY 2: CRISPR-EDITED WHEAT TO REDUCE GLUTEN ALLERGENS

CRISPR has been employed by researchers to specifically search out and deactivate certain genes that cause gluten proteins implicated in provoking celiac disease. This process creates wheat that has much lower levels of gluten. Although the feature of CRISPR-edited wheat is broadly similar to traditional wheat in terms of baking quality and yield, early studies indicate that it can be safer for those with gluten sensitivities. Wheat is among the most prominent food crops, but it does come with health risks to non-celiac gluten/no-grain tolerants. This protein is present in wheat, barley, and rye; but those suffering from celiac disease could develop serious digestive complaints as well as long-term health consequences [29]. This trait has not been successfully changed in traditional wheat breeding so far, but genome editing opens up new possibilities for reducing gluten content. Although the improved wheat seems to be beneficial, we still require more research which needs to be done in order to confirm its antiallergenic effects on humans and make sure that removing gluten won't cause other side effects, such as affecting nutritional values or digestibility. Moreover, easy availability of gluten-reduced wheat may have exciting implications for anyone who has any form of allergy to gliadins or celiac disease.

7.6.3 CASE STUDY 3: POTATO PROTEIN CONTENT INCREASED WITH FE-SUL APPLICATION

Acrylamide is a chemical that can form in some foods–mainly plant-based food such as potato products, grain products, and coffee–during high-temperature cooking processes like frying, and it is closely related chemically to estragole, a potential carcinogen with mutagenic effects that has led some health experts to express concerns about its presence in food. By editing the genes that help asparagine form, genome editing can be used to lower acrylamide levels in potatoes [30]. A potato variety developed through genome editing can help to limit the health risks associated with acrylamide consumption in items such as French fries and chips, both popular foods around the world.

The stakes for health effects in humans are nominal, or so is the expectation; acrylamide represents an ideal toxicant that can be thwarted simply by limiting dietary exposure to potentially harmful compounds.

However, like other genome-edited crops, full safety assessments and long-term studies are mandatory before these variants can hit the market to make sure that modifications made do not bring on health risks or distort its nutritional value [31].

7.6.4 *CASE STUDY 4: OMEGA-3 FATTY ACID–RICH SOYBEANS*

Omega-3's are essential nutrients and well documented in the areas of heart health, brain function, and inflammation regulation. Omega-3 fatty acids are mainly found in fish and some plants, but they cannot be obtained completely from diet only due to the fact that improper intake may either result in a lower echelon of performance or an exacerbation through tissue. Researchers have also used genome editing to produce oilseed from soybeans with increased quantities of omega-3 fatty acids, which could serve as a new source for the plant-derived nutrients. In the study, researchers employed CRISPR to alter the pathways of fatty acid synthesis in soybeans that can elevate omega-3-fatty acid content [32]. These unsaturated fats are known to reduce heart disease risk, and the genome-edited soybeans could provide a sustainable, economically viable source of these nutritious oils that have wide-ranging health benefits, especially in lowering cardiovascular diseases. Over more than 20 years, the health benefits of omega-3 fatty acids have been extensively researched (~100 clinical studies), but development and introduction of genome-edited soybeans raise questions about consumer acceptance as well as regulatory oversight needed to ensure such a source could meet all required criteria for safe consumption [33]. Future research should also address whether omega-3 fatty acids in these modified soybeans are bioavailable as compared to those sourced from wild fish or type of purified marine oil.

The application of genome-edited plants in human health is promising, especially from the perspective that enhanced nutrition (nutritionally improved) and reduced allergenicity and toxicity by foodborne toxins may increase food safety. Golden Rice, CRISPR-edited wheat, low-acrylamide potatoes, and omega-3-enhanced soybeans are case studies showing how genome editing can improve the nutritional content of staple crops. Yet, these latest advances also highlight the imperative for continuous research to implement rigorous safety assessments and appropriate regulatory frameworks that are long needed in order to guarantee that all genome-edited plants present no unintended health risks when consumed by humans [34].

7.7 CONCLUSION AND FUTURE DIRECTIONS

Plant genome editing, especially the CRISPR-Cas9 technique, is one of the most promising tools to contribute to food security, nutrition, and sustainability. Yet, it also raises serious biosafety issues and possible health risks. Genome-edited crops like Golden Rice and low-gluten wheat that are being developed to overcome serious health issues such as vitamin deficiencies and food allergies require long-term safety assessments. The unintended consequences of alteration and other uncertainties, including gene flow to wild species and ecological impacts from the use of Golden Rice, may need careful regulation. To address ethical issues and allow room for public confidence, transparent labeling is vital, along with strong safety protocols.

The future of genome editing efforts must be built on a foundation of strong biosafety frameworks; there should not arise from the process any GMOs that might have unintended health or environmental risks. More collaboration between scientists and policymakers, as well as the general public, is needed for recommendations that meet innovation with safety. Given their potential in breeding genes that promote chronic disease, interdisciplinary research can be accelerated to study significant and long-term effects of a genome-edited plant on human health as technology advances. Genome editing has the potential to revolutionize our world over time, permitting broad public deployment through complex biosafety and ethical/regulatory constraints with appropriate responsibility in the promotion of human welfare.

KEYWORDS

- **biodiversity**
- **biosafety issues**
- **biotechnology**
- **CRISPR-Cas9 technique**
- **genetic mutations**
- **genome editing**

REFERENCES

1. Guru, R. K., Ganpatrao, A. S., Madhao, A. P., Pradhan, R., Mohanty, A., Panigrahi, K. K., Tarai, R. K., & Khatoon, B. (2024). Biosafety and biosecurity concerns associated with plant genome editing. In *Genome Editing and Global Food Security* (pp. 236–274). Routledge.
2. Sabitha, N. (2024). Biosafety issues of genetically modified crops. https://agritechmagazine.com (accessed on 5th January 2026).
3. Singh, B., & Kaunert, C. (2024). Adventure in high altitude of mountainous topographies and health impacts: Lensing tourism sustainability via reducing ecological and sociocultural footprint and health emergency and medical assistance management. In G. Meraj, S. Hashimoto, & P. Kumar (Eds.), *Navigating Natural Hazards in Mountainous Topographies. Disaster Risk Reduction* (p. 15). Cham: Springer. https://doi.org/10.1007/978-3-031-65862-4_15 (accessed on 5th January 2026).
4. Munawar, N., Faheem, M., Niamat, A., Munir, A., Khan, S. H., Zahoor, M. K., Aslam, A., & Ahmad, A. (2024). Regulatory, ethical, social, and biosafety concerns in genome-edited horticultural crops. In *CRISPRized Horticulture Crops* (pp. 421–438). Academic Press.
5. Hassan, N., Ali, H., Hassan, F., Ali, H., & Tawab, A. (2024). Biosafety and biosecurity consideration in CRISPR-Cas. In *CRISPRized Horticulture Crops* (pp. 507–520). Academic Press.
6. Singh, B., Singh, A., Kaunert, C., Arora, M. K., Lal, S., & Ravesangar, K. (2025). Unleashing the ethical and legal implications of e-business revolution: Consumer privacy and security concerns in phase of digital disruption. In H. Taherdoost, G. Drazenovic, M. Madanchian, I. Khan, & O. Arshi (Eds.), *Business Transformation in The Era of Digital Disruption* (pp. 157–180). IGI Global. https://doi.org/10.4018/979-8-3693-7056-8.ch006 (accessed on 5th January 2026).
7. Parvathy, S. T. (2024). Genome editing: A safe alternative to genetic engineering of crops. In *Genetic Engineering of Crop Plants for Food and Health Security: Volume 2* (pp. 327–372). Singapore: Springer Nature Singapore.
8. Yadav, V., Pal, D., & Poonia, A. K. (2024). A study on genetically engineered foods: Need, benefits, risk, and current knowledge. *Cell Biochemistry and Biophysics, 82*(3), 1931–1946.
9. McFadden, B. R., Rumble, J. N., Stofer, K. A., & Folta, K. M. (2024). US public opinion about the safety of gene editing in the agriculture and medical fields and the amount of evidence needed to improve opinions. *Frontiers in Bioengineering and Biotechnology, 12*, 1340398.
10. Singh, B., & Kaunert, C. (2025). Innovative business models in digital and virtual reality for the gig economy: Space-age sustainable development projecting smart business practices. In H. Taherdoost, G. Drazenovic, M. Madanchian, I. Khan, & O. Arshi (Eds.), *Business Transformation in the Era of Digital Disruption* (pp. 127–156). IGI Global. https://doi.org/10.4018/979-8-3693-7056-8.ch005 (accessed on 5th January 2026).
11. Cetiner, S. (2024). The impact of regulatory oversight on the development and adoption of plant biotechnology. In *Book of Proceedings* (p. 173).

12. Kumawat, T., Agarwal, A., Saxena, S., & Arora, S. (2024). An analysis of global policies and regulation on genome editing in plants. In *Gene Editing in Plants: CRISPR-Cas and Its Applications* (pp. 775–793).
13. Singh, B., Malviya, R., & Kaunert, C. (2024). Elevating workplace sustainability for employees lensing mental health advancements: Runway for future ready healthcare services projecting SDG 3 (good health and well-being). In E. Shaikh (Ed.), *Impact of Corporate Social Responsibility on Employee Wellbeing* (pp. 285–310). IGI Global. https://doi.org/10.4018/979-8-3693-3470-6.ch014 (accessed on 5th January 2026).
14. Sabat, M., & Tripathy, A. (2024). Genetically modified and gene-edited food crops: Recent status and future prospects. In *Food Production, Diversity, and Safety Under Climate Change* (pp. 211–222).
15. Singh, B., Kaunert, C., & Malviya, R. (2024). Featuring machine learning in municipal solid waste management: Technological advancements in waste sorting and processing ropeway for SDG 13 (climate action). In R. Mehra, P. Kumar, & B. Hung (Eds.), *Municipal Solid Waste Management and Recycling Technologies* (pp. 343–366). IGI Global. https://doi.org/10.4018/979-8-3693-4054-7.ch013 (accessed on 5th January 2026).
16. Munawar, N., Ahsan, K., & Ahmad, A. (2024). CRISPR-edited plants' social, ethical, policy, and governance issues. In *Global Regulatory Outlook for CRISPRized Plants* (pp. 367–396). Academic Press.
17. Adeel, M., & Jones, M. G. (2024). Challenges and prospects in the regulation of CRISPR-edited crops. *Global Regulatory Outlook for CRISPRized Plants*, 447–459.
18. Singh, B., Kaunert, C., & Jermsittiparsert, K. (2024). Environment-biodiversity protection and SDG 15 (life on land): Legal landscapes and probable resolutions. In B. Singh, C. Kaunert, K. Vig, & S. Dutta (Eds.), *Maintaining a Sustainable World in the Nexus of Environmental Science and AI* (pp. 259–282). IGI Global. https://doi.org/10.4018/979-8-3693-6336-2.ch010 (accessed on 5th January 2026).
19. Catherine, K. N., Mugiira, B. R., & Muchiri, N. J. (2024). Public perception of genetically modified organisms and the implementation of biosafety measures in Kenya. *Advances in Agriculture, 2024*(1), 5544617.
20. Weldemichael, M. Y., Gebremedhn, H. M., & Teklu, T. H. (2024). Advances in genome editing and future prospects for *Sorghum* improvement: A review. *Plant Gene*, 100464.
21. Singh, B., & Kaunert, C. (2024). Transforming smart city and smart sustainable tourism projecting artificial intelligence and IoT: Immersive potential for revolutionizing urban living and enhancing travel experience. In K. Jermsittiparsert & P. Suanpang (Eds.), *Technological and Managerial Approaches to Fostering Sustainable Travel* (pp. 55–80). IGI Global. https://doi.org/10.4018/979-8-3693-8115-1.ch003 (accessed on 5th January 2026).
22. Shan, P. (2024). Application and prospect of genetically modified technology in agriculture. *GMO Biosafety Research, 15.*
23. Raghav, A., Singh, B., Jermsittiparsert, K., Raghav, R., & Yadav, U. (2024). Artificial intelligence in environmental and climate changes: A sustainable future. In B. Singh, C. Kaunert, K. Vig, & S. Dutta (Eds.), *Maintaining a Sustainable World in the Nexus of Environmental Science and AI* (pp. 485–506). IGI Global. https://doi.org/10.4018/979-8-3693-6336-2.ch019 (accessed on 5th January 2026).

24. Ahmad, A., Arif, A., & Munir, A. (2024). CRISPR technology commercialization and biosafety. In *Global Regulatory Outlook for CRISPRized Plants* (pp. 461–514). Academic Press.
25. Kaushik, T. K., Ravish, Singh, A., Raghav, A., Singh, B., Arora, M. K., & Lal, S. (2024). Untapping artificial intelligence in environmental and sustainable progression: Projecting legal and ethical considerations. In B. Singh, C. Kaunert, K. Vig, & S. Dutta (Eds.), *Maintaining a Sustainable World in the Nexus of Environmental Science and AI* (pp. 109–130). IGI Global. https://doi.org/10.4018/979-8-3693-6336-2.ch005 (accessed on 5th January 2026).
26. Irfan, M., Majeed, H., Iftikhar, T., & Ravi, P. K. (2024). A review on molecular scissoring with CRISPR/Cas9 genome editing technology. *Toxicology Research, 13*(4).
27. Singh, B., Kaunert, C., Malviya, R., Lal, S., & Arora, M. K. (2024). Scrutinizing consumer sentiment on social media and data-driven decisions for business insights: Fusion of artificial intelligence (AI) and business intelligence (BI) foster sustainable growth. In A. Natarajan, M. Galety, C. Iwendi, D. Das, & A. Shankar (Eds.), *Intersection of AI and Business Intelligence in Data-Driven Decision-Making* (pp. 183–210). IGI Global. https://doi.org/10.4018/979-8-3693-5288-5.ch007 (accessed on 5th January 2026).
28. Abd-Elsalam, K. A., & Ahmad, A. (2024). Regulations of genome edited plants: Notes from editors. In *Global Regulatory Outlook for CRISPRized Plants* (pp. 1–20). Academic Press.
29. Singh, B., Kaunert, C., Jermsittiparsert, K., Lal, S., Arora, M. K., & Raghav, A. (2024). Ultimate deep learning acclimatization with artificial intelligence drones for environment shield and climate change: Lifting sustainability towards fostering SDG 13. In B. Singh, C. Kaunert, K. Vig, & S. Dutta (Eds.), *Maintaining a Sustainable World in the Nexus of Environmental Science and AI* (pp. 59–80). IGI Global. https://doi.org/10.4018/979-8-3693-6336-2.ch003 (accessed on 5th January 2026).
30. Devi, M., Tushir, N., Parashar, N. C., & Markandeshwar, M. (2024). The impact of biotechnology on food production: A paradigm. *Prof. Ashish Agrawal*, 85.
31. Singh, B., Kaunert, C., & Gautam, R. (2024). Harnessing machine learning intelligence against cyber threats: Navigating legal and ethical challenges in global cyber investigations. In F. Özsungur (Ed.), *Empowering Entrepreneurial Mindsets with AI* (pp. 319–352). IGI Global. https://doi.org/10.4018/979-8-3693-7658-4.ch016 (accessed on 5th January 2026).
32. Cummings, C., Selfa, T., Lindberg, S., & Bain, C. (2024). Identifying public trust building priorities of gene editing in agriculture and food. *Agriculture and Human Values, 41*(1), 47–60.
33. Clark, L. F., & Hobbs, J. E. (2024). How are gene editing technologies regulated in the agrifood system? In *International Regulation of Gene Editing Technologies in Crops: Current Status and Future Trends* (pp. 31–59). Cham: Springer Nature Switzerland.
34. Brookes, G., & Smyth, S. J. (2024). Risk-appropriate regulations for gene-editing technologies. *GM Crops & Food, 15*(1), 1–4.

CHAPTER 8

Gene Editing and Human Rights: Legal Implications of CRISPR Technology

FATIMA ZAHRA OUARIACH[1] and MANMEET KAUR ARORA[2]

[1]*Abdelmalek Essaadi University, Tétouan, Morocco*

[2]*Sharda University, Greater Noida, Uttar Pradesh, India*

*Corresponding author

ABSTRACT

The chapter examines the complex terrain of cloning and genetic manipulation, focusing on key accomplishments as well as ethical questions and societal consequences. Major case studies, such as that of Dolly the sheep and therapeutic cloning in medical research, demonstrate potential benefits to medicine, agriculture, and conservation. This leads to discussions about the moral status of clones, informed consent issues, and psychological well-being effects on clones, along with long-term impacts for biodiversity. It investigates the role of public perceptions in shaping public policies, calling for transparent communication and engagement to ensure trust in scientific progress. The chapter also considers the need for holistic legal regimes that can regulate cloning appropriately, while at the same time ensuring fair access to technology. It also presents an example of the future of cloning and genetic manipulation, for which an answer is given on whether we should seek to expand existing tools' applications. It reaffirms the importance of proactive solutions to ethical dilemmas in learned behavior and responsible actions within biotechnology.

Genetic Editing for Healthcare: Advances in Disease Management.
Bhupinder Singh, Rishabha Malviya, Sahil Lal, and Sathvik Belagodu Sridhar (Eds.)

DOI: 10.1201/9781779643827-8

8.1 INTRODUCTION

Cloning and genetic manipulation also represent some of the cornerstones of modern biotechnology, which is a key area in development that can revolutionize medicine, agriculture, and genetics. "Clone"—this term is used to create genetically identical organisms or cells. This process sometimes occurs naturally, as it does in identical twins and clones. It can also be achieved artificially using techniques such as somatic cell nuclear transfer (SCNT), which was used to create Dolly the sheep in 1996. There are multiple types of cloning, but the primary focus lies in two: reproductive and therapeutic cloning. Reproductive clones aspire to give rise to an entire organism for breeding or further experimentation with recombinant DNA technology (after sequence-specific genome editing has been applied), while those who perform research on cells which resemble humans often wish that they would one day be able to create such animals using either genetic engineering alone or biological means [1].

The definition of genetic manipulation is that it covers many techniques used to alter the characteristics of an individual organism by modifying its DNA [2]. Methods such as CRISPR-Cas9 empower pinpoint DNA sequence modifying that can be used to correct genetic diseases, improve the strength of crops, or develop gene therapies for multiple illnesses. While these technologies promise many benefits, they also raise serious ethical and legal issues that society will need to grapple with [3].

8.2 THE LEGAL AND ETHICAL PERSPECTIVE

With the evolution of cloning and genetic manipulation technologies, such robust legal and ethical frameworks are ever more necessary. These technologies are about more than just scientific curiosity: they touch on some of the most basic human rights, questions of identity, and moral foundations [4].

8.3 HUMAN RIGHTS AND AUTONOMY

Here, autonomy and human rights violation is one of the central ethico-moral lines in cloning. Critics say it opens the door to a world where human life is treated like any other good, and people are essentially manufactured in response to certain wants or needs—whether for company or as an organ donor. This is the view that overlaps with Kantian ethics, which holds that we should never treat people only as means to an end but always also as ends in themselves. These ethical concerns are also playing out in the legal landscape. Several countries have banned human reproductive cloning, some in response to anxieties that it violates the sanctity of being a person [5]. Take the United Nations, for example — it has established principles Real World Impacts on Identity and Society [6].

Challenging Issues with Identity and Social Acceptance for clone species: Because they are exact replicas of their donors, the concerns around clones turn mainly on whether or not to treat them as individuals with independent rights [7]. Would they be considered equals in the eyes of society? Or would they be discriminated against, boiled down to nothing more than the sum of their genetic parts? But these are the questions that raise a fundamental issue, which is what legal measures, if any, can be pursued to ensure clones themselves should also have privileges of rights and dignity like all other members of society. Secondly, there is obvious discomfort about the specter of eugenics, an institutional process aimed at producing offspring with preferred traits. The real "worst case" scenario may involve cloning technologies misused to make place a run on your standard old designer babies with the added kicker that those same parents could then argue their descendants are exactly where they belong, while everyone else is left sterilized or dead. Where that might seem to have problems with ethics around diversity and privileging one attribute over another [8]. Outlawing human cloning in an attempt to safeguard individual rights pertaining to such biotechnologies. The first kind of account, usually grounded in freedom and autonomy considerations (*see* Baylis, 1997), claims that reproductive cloning is unethical because it imposes a predetermined genetic identity on an as yet non-existent person [9].

8.4 THERAPEUTIC CLONING: ANOTHER POINT OF VIEW

Although reproductive cloning involves some serious moral questions, therapeutic cloning gives a more complicated one. This is a type of cloning where embryos are created for research to extract stem cells that could be useful in treating diseases and other ailments, but without ever implanting the embryo into a womb [10]. In addition, supporters of therapeutic cloning contend that the practice offers promising avenues toward achieving major achievements in regenerative medicine and disease treatment without running afoul such a long list of ethical roadblocks as reproductive cloning does [11]. Still, therapeutic cloning raises its own ethical issues. The use of embryos for research brings up moral issues as to when life begins and whether it is acceptable (in the name of medical progress) to destroy what could be potential human beings. Medical-legal frameworks must walk a fine line between protecting the privacy of patient data and promoting scientific advances [12].

8.5 UNDERPINNINGS OF CLONING TECHNOLOGIES: TYPES OF CLONING

In short, cloning is a diverse process in biology, and there are several types of it depending on what you need to achieve or the type that requires being cloned. There are three main types of cloning: gene cloning, reproductive cloning (position your cursor over the image for more info), and therapeutic [11].

8.6 GENE CLONING

Gene cloning: Gene or DNA cloning is the process of obtaining copies of specific DNA pieces and genes. The method is foundational to much of molecular biology and indispensable in modern applications such as gene therapy, genetic engineering, and protein production [12]. In practice, gene cloning typically involves the following steps:

1. **DNA Isolation:** The first step is to remove the relevant fragment of DNA from the genome. Methods for achieving this include, but are not restricted to, the use of restriction enzyme digestion, which cleaves DNA at defined portions.
2. **Insertion into a Vector:** Once extracted, the DNA fragment must be inserted into a vector. Most vectors (such as plasmids and bacteriophages) are polymers of cloning vehicles, although any molecule, such as an entire chromosome, can also carry foreign DNA. Vectors are important because they transport the foreign DNA into host cells and make them replicated and expressed [13].
3. **Transformation:** It is the process via which a vector that carries the DNA fragment is incorporated into host cells (usually *E. coli* bacteria). The host cells replicate, and multiple copies of the inserted DNA are created [2].

4. **Screening and Selection:** After transformation, researchers use selective media to find a successfully transformed cell where your gene of interest is cloned. In most of these cases, specific antibiotic resistance markers are contained in the vector [3].
5. **Analysis or Amplification:** The cloned gene can be further amplified for analysis (PCR, sanger sequence) and downstream applications.
6. **Applications of Gene Cloning:** Gene cloning is the foundation of modern biotechnology as it enables large-scale production of insulin, growth hormone, and other therapeutic proteins. It is also essential in studying genetic research and allows scientists to study the functioning of genes as well as regulation [4].

8.7 REPRODUCTIVE CLONING

The goal of reproductive cloning is to create a genetically identical organism [5]. This kind of cloning is quite controversial and has the ability to be applied in conservation biology, though. SCNT is the most common method for reproductive cloning. The SCNT process entails several essential steps.

It literally shrinks back into a non-specialized ball of nuclei, which is sucked out to create a pool of DNA material from one of the individual cloning targets. Inside the nucleus of this cell resides the genetic material for that organism [6].

1. **Oocyte Retrieval:** An oocyte (an unfertilized egg cell) is collected from a donor organism. This egg is then enucleated, meaning its nucleus is extracted to make room for the donor DNA.
2. **Nuclear Transfer:** The nucleus of the donor's somatic cell is inserted into the enucleated egg cell. This reconstructed egg now contains the genetic material of the donor organism [7].
3. **Embryo Culture:** The egg containing the donor DNA is stimulated—either chemically or electrically—to begin development, mimicking the natural fertilization process. This leads to the formation of an embryo.
4. **Embryo Implantation:** Once the embryo has developed to an appropriate stage, it is implanted into a surrogate mother, where it continues to grow into a fully formed organism [8].

Reproductive cloning has been demonstrated successfully in several species, such as sheep (*e.g.*, Dolly), cattle, and horses. However, it is associated with high failure rates and significant ethical concerns, particularly regarding animal welfare. Many clones experience health issues, raising questions about the long-term viability and morality of the process.

8.8 THERAPEUTIC CLONING

8.8.1 HUMAN EMBRYO CLONING FOR THERAPEUTIC USE

This is the use of SCNT to produce cloned embryos with a view toward deriving stem cells from them, not developing an entire organism. All this bodes extremely well for the promise of regenerative medicine and tissue engineering [9]. In its goals, the therapeutic cloning process closely resembles that of reproductive cloning, even explicitly keeping some steps, but offers opposite means.

8.8.2 CREATING THE EMBRYO

Just like with reproductive cloning, therapeutic cloning starts by fertilizing an egg (via SCNT) with a nucleus from any somatic cell [10].

Instead of letting the embryo grow into a full organism, scientists extract stem cells from an embryo at the blastocyst stage (a few days after fertilization). This property is known as pluripotent (because they can differentiate into just about any cell type in the body).

8.8.3 CELL DIFFERENTIATION

After isolation, these stem cells can be cultured in conditions that help them attain the cell types (e.g., neurons or cardiac) needed for transplantation or research [11].

Therapeutic cloning is important in the context of treatment for different diseases (Parkinson's disease, diabetes, and spinal cord injuries) because it provides patient-specific cells that would allow marked reduction in rejections following transplants. Yet, it raises issues of whether using and destroying embryos is ethically appropriate [12].

8.9 RESEARCH AND DEVELOPMENT TREND/MAIN APPROACH

Cloning technologies continue to develop further, with current improvements and novel applications still being investigated [13].

8.9.1 ADVANCES IN GENE CLONING

Advances in gene cloning methods: The recent advances in the molecular biology field include CRISPR-Cas9 technology for editing genes, which allows directed genome modifications. The technology allows very precise corrections to the genome for genetic defects that cause hereditary diseases. In this context, the development of synthetic biology (a related field to gene cloning) has provided a powerful new means for constructing novel biological parts and systems [1]. This enables researchers to create synthetic genes that can perform various tasks in several organisms, opening doors for novel therapeutics and biotechnological applications [2].

8.9.2 REPRODUCTIVE CLONING: AN UPDATE

Reproductive cloning continues to be shrouded in controversy due to its ethical implications, but research for refining techniques such as SCNT does not cease, so that success rates and outcomes regarding cloned animals can improve. We also need to look at improving embryo health and minimizing the number of genetic abnormalities characteristic of most clones by learning more about epigenetic drivers in development. Reproductive cloning could also prove to be useful for conservation efforts by creating genetic copies of endangered species [3].

8.9.3 THERAPEUTIC CLONING ADVANCES

Current research in therapeutic cloning aims to optimize techniques involved with the derivation of stem cells from cloned embryos and also concentrates on addressing many ethical considerations associated with embryo use [4]. One potential alternative to human embryonic stem cells is pluripotent stem cell lines derived from other sources (e.g., induced pluripotent stem [iPS] cells), generated by reprogramming adult somatic tissues. The discovery of iPSCs has changed the way we think about stem cell biology and regenerative medicine; with them, scientists have finally found a feasible source that can replace pluripotent cells for research and therapeutic development without destroying an embryo [5].

8.10 LEGAL REGIMES ON CLONING: RULES AND REGULATIONS ON CLONING GLOBALLY

8.10.1 INTERNATIONAL CLONING REGULATION

An Ever-Changing Patchwork Pregnant Elephant Syndrome. No binding international agreement exists on human cloning, which falls under the regulation of devices for reproduction, so it has been regulated—but no such comprehensive global treaty covers jurisdiction over cloning—and there are current declarations not coming from those that desire moral and ethical governance [14].

8.10.2 ROLE OF THE UNITED NATIONS

In the world, there is a lot of talk of cloning, and the United Nations (UN) has largely impacted how this discourse goes internationally. In 2005, the UN General Assembly sponsored a non-binding Declaration on Human Cloning, which asked member states to prohibit all forms of human cloning incompatible with human dignity and protection of life (or "human rights"). This declaration demonstrates an international consensus on the ethical aspects of reproductive cloning, mainly with respect to personal integrity and identity [15].

International calls for a convention to ban reproductive cloning led the UN to become involved in 2001, following pressure from France and Germany. Yet, the declaration is not legally binding for member states and is simply an orientation framework when writing national laws. This non-binding characteristic further signifies that the countries are advised to adopt prohibitive measures; however, they have ample discretion as far as their jurisdiction is concerned with respect to regulation on cloning in it [16].

8.10.3 GUIDELINES ARE NON-BINDING

Since international guidelines are not legally binding, this status makes it difficult to apply the same strict regulations globally for cloning. In light of this, different countries have pursued their healthcare guidance because the approaches differ depending on cultural and ethical views, as well as scientific reasoning [17]. For example, some countries have prohibited all forms of human cloning, whereas others allow therapeutic cloning under several conditions. Global standards with consequences are non-existent, allowing for "cloning tourism" to flourish because researchers can escape stricter regulation by traveling between countries. Given this backdrop, our regulators globally should seriously consider the promotion of a more unified international regulatory framework to safeguard ethics and safety-related concerns pertaining to duplicating technologies [18, 22].

8.10.4 NATIONAL LAWS AND APPROACHES

The laws in connection with national cloning differ depending on the country, and these differences reflect a broad array of cultural values and ethical tools [19]. A few countries have passed legislation that specifically mentions reproductive and therapeutic cloning, but others only refer to a ban on any form of human intervention in the early embryo [20].

8.10.5 NATIONAL APPROACHES ON THE GROUND

1. **Human Fertilization and Embryology Act 1990:** The UK has established a clear regulatory framework through the Human Fertilization and Embryology Act of 1990, which allows therapeutic cloning under stringent licensing conditions while banning reproductive or anything other than experimental cloning [24]. The HFEA is the United Kingdom's independent regulator which oversees licensing for human embryo research, making sure it takes place within a strict ethical framework [21, 25].
2. **United States:** There is no commercial human cloning ban in the U.S., but a few states restrict or prohibit reproductive cloning. Researchers work with a complex legal code, as the Dickey-Wicker Amendment bans federal funds for research that destroys or harms human embryos [26].

8.11 NATIONAL LAWS AND APPROACHES

The laws in connection with national cloning differ depending on the country, and these differences reflect a broad array of cultural values and ethical tools [23]. A few countries have passed legislation that specifically mentions reproductive and therapeutic cloning, but others only refer to a ban on any form of human intervention in the early embryo.

8.11.1 GUIDELINES ARE NON-BINDING

Also, it is specified that the recommendations from international bodies are not lawfully enforceable. This status means it would be hard to sternly impose similar universal restrictions for cloning. In this context, diverse nations have followed these medical directives since cultures and ethics differ, as well as for scientific reasons. For example, while some countries have enacted complete bans on all forms of human cloning, others permit therapeutic cloning under various conditions [27]. The lack of global standards, coupled with little to no repercussions, has become an ideal environment for a new type of "cloning tourism"--the ability to hop borders and perform research in any country where they face less scrutiny. Against this context, it seems opportune to have a serious thought of promoting an all-encompassing international regulatory standard framework by our respective regulators across the world so as to protect interests in respect of ethics and safety issues specific for replicating technologies [28].

8.11.2 NATIONAL LAWS AND APPROACHES

The norms associated with national cloning vary from country to country and can signify the different national individual values as well as permissive additions and restrictions. Some nations have enacted laws that refer to reproductive and therapeutic cloning explicitly, while others address a prohibition on any form of human intervention in the early embryo [29].

8.11.3 NATION'S APPROACH

1. **UK Human Fertilization and Embryology Act 1990:** The UK has in place the principle of positive regulation by means of its Human Fertilization and Embryology Act (HFEA). Therapeutic cloning is permitted under very strict licensing conditions, whereas reproductive or any other than pilot cloning is explicitly forbidden. The HFEA is the body that licenses and monitors all UK research conducted on embryos [30].

2. **USA:** Human cloning is not entirely banned in the U.S. No federal ban on human cloning exists, and only some states restrict or prohibit reproductive cloning. The Iron Balls of the Law: The Dickey-Wicker Amendment prohibits federal funds for "research in which a human embryo or embryos are destroyed, discarded, or knowingly subjected to risk."

8.12 CASE STUDY: UK HUMAN FERTILIZATION AND EMBRYOLOGY ACT (HFEA)

The HFEA 1990 is a vital piece of UK law to safely oversee assisted reproductive technologies, chiefly cloning and embryo research. The purpose of this Act was to prevent misuse in a new field where developments are being done with human embryos. This legislation extends to establish a complete legislative structure for the regulation of fertility procedures, embryo research, and human gametes being utilized [31].

8.12.1 *HISTORICAL CONTEXT*

The HFEA followed a period of public debate and ethical discussions on issues raised by reproductive technologies, particularly after the birth in 1978 of Louise Brown, the world's first "test-tube baby." The increasingly rapid development of assisted reproductive technologies sparked various issues about the morality of playing with human life, even before birth. In reply to this, the UK government formed a committee that was led by Dame Mary Warnock, focusing on these issues and leading up to HFEA creation [32].

8.12.2 *KEY PROVISIONS OF THE HFEA (HUMAN FERTILIZATION AND EMBRYOLOGY AUTHORITY)*

1. **Embryo Research:** The HFEA permits human embryo research as under regulation. It is illegal to conduct research on older embryos, because the law was designed to cover only IVF. This restriction is designed to strike a balance between the potential of scientific inquiry and ethical concerns about embryo status [33].
2. **Licensing:** All fertility clinics and other research establishments that engage in activities with early embryos must receive a license

from the HFEA. This in turn ensures that practices are conducted according to accepted ethics and safety guidelines.

3. **Reproductive Cloning:** The HFEA stipulates that reproductive cloning is banned due to ethical objections in relation to the birth of identical human beings. The ban is based on the argument that cloning dehumanizes and tends to regard people as interchangeable.
4. **Consent Requirements:** The act places an increased emphasis on the informed consent of individuals undergoing fertility treatments or participating in research that involves their own embryos or gametes. This clause is here to protect the patient's rights and to ensure that they understand what their decision could imply.
5. **Donor Gametes:** Donors must provide written consent, and they are tested for genetic conditions in the UK by the HFEA. This is done to ensure that they do not threaten the health of future offspring.

8.12.3 IMPACT ON CLONING RESEARCH

The HFEA has been influential in the regulation of cloning research in this country, striking a balance between scientific progress and ethical concerns. Therapeutic cloning is allowed under very tightly controlled guidelines, but reproductive cloning remains illegal. This separation could offer a way for researchers to investigate the medical potential of stem cell therapies without stumbling into ethical hot water over human cloning [33].

8.12.4 ETHICAL CONSIDERATIONS

Values for human dignity, autonomy, and the moral status of embryos are not merely arbitrary preferences without foundation: those principles also underpin a range of agreed social norms which most members accept within our society, although some variation will remain. The act aims at addressing a number of difficult ethical issues related to reproductive cloning and embryo research by banning the former, whilst controlling both through regulations.

8.13 CHALLENGES AND CRITICISMS

However, having laid out such an extensive framework, the HFEA must balance several competing priorities. Countries are emerging from conflicts, while other countries continue to be unstable such as:

1. **Impact on Public Perceptions:** One issue lies here in that many of the utility identification arguments work. However, there are still ongoing debates within society regarding cloning and genetic manipulation, to name but a few, which can play an important role due to the current mode for debate. This would affect sentiments towards the HFEA, regarding whether or not it is doing enough to preserve those standards [34].
2. **Technology:** Biotechnology develops rapidly, which may overtake current standards and hence demand constant strategic alignment.
3. **Vagaries in the Law:** Critics say that some provisions of the act could be vague or subject to interpretation, thereby leading to discrepancies in enforcement.

8.14 CASE STUDY: U.S. STATE-LEVEL REGULATIONS

Contrary to nations that have adopted comprehensive national legislation regarding cloning and reproductive technologies (e.g., UK), the current state of regulation in this country is characterized by a patchwork of laws enacted at the individual state level, including those pertaining to certain of these practices involving transgenic animals. Having a piecemeal system creates problems for researchers, patients, and policymakers.

8.14.1 CHALLENGES FACED BY STATES DUE TO LEGALIZATION

1. **Conflicts Among State Laws:** In states that may have fairly different sets of laws, researchers could face unnecessary legal challenges when doing work across state lines because there is no universal standard (such as which cloning practices are acceptable).
2. **Uncertain Legal Environments:** There may be limited or unclear definitions in state statutes that create confusion about whether

specific cloning-related practices are considered permissible or impermissible.

a. **Enforcement Application:** Lack of uniformity in enforcement methods from state to state can lead to an unleveled playing field with laws being applied differently by courts or regulators.

b. **Changes in Public Opinion:** Biotechnology is not as leftist/ rightist issue technically, making its enforcement susceptible to regulatory shifts that reflect local morality.

8.14.2 ETHICAL CONSIDERATIONS

There are also mandates at the state level that themselves impose ethical questions:

1. **Access to Treatments:** Indeed, disparities in state regulations can have major consequences on access to cutting-edge treatments, for instance, stem cell therapies, based upon where those patients actually live.
2. **Patient Rights:** There is variability among states that may impact the rights of patients with respect to informed consent processes related to their participation in research using embryos or gametes.
3. **Research Opportunities:** States with lax regulations may draw researchers who wish to take advantage of more lenient environments for conducting novel research, whereas those that have passed strict laws could impede scientific advancement.

8.14.3 LITIGATIONS AND LEGAL LACUNAS

The pace of technological developments in the field of biotechnology is unmatched and ever-expanding; however, existing cloned-based legal frameworks bleed loopholes because of lacks in legislatures and inconsistencies within laws. These questions make it more difficult to properly control these new technologies without maintaining ethical standards.

8.14.3.1 DEFINING LEGAL GAPS

Legal gaps result from inadequate statutes to deal with new issues or relationships generated by a variety of new technologies—including cloning and related activities—leaving uncertain where the lines are drawn between permitted behavior and prohibited activity.

- **Types of Legal Gaps:**

1. **Full Absence:** Circumstances where there are no laws regarding certain activities in cloning.
2. **Biotechnology Practice Ambiguities:** Towards different aspects, lack of clarity/specificity in law already existing.
3. **Obsolete Regulations:** Legal rules that have been made in advance of the key technological transformations may no longer be well-suited for a host of compelling scientific fields such as genetics or reproductive technology.

8.14.3.2 SOME GAP FROM LEGAL PERSPECTIVE RELATED WITH CLONING

1. **Absence of Nationwide Legislation:** Gaps in legislative coverage at both the national and state levels result from the failure to enact comprehensive federal legislation regulating human cloning.

   Many laws lack clear definitions of human embryo cloning and stem cell research, resulting in uncertain legal parameters and ambiguity regarding what practices are allowed?

2. **Technology Moving Faster Than Legislation:** Advances within the realm of biotechnology often give rise to novel procedures before complementary legal frameworks are established—leaving large gaps in ethical oversight that are not covered by current laws.

   Popular read-ahead policies exclude ongoing ethical review in response to new biotechnologies, potentially opening the door to unapproved experimentation or commercialization efforts with associated risks.

3. **Interstate Conflict:** Since state cadaver laws vary, confusion arises in studies conducted across states regarding compliance with different regulatory requirements for cloning practices [35].

8.14.4 IMPLICATIONS OF LEGAL GAPS

1. **Lack of Regulation:** Legal gaps cause researchers to be uncertain about what they can and cannot do, which in turn causes financial investors to shy away from entering into innovative biotechnologies for fear of compliance risk.
2. **Moral Complexities:** Failures in these processes lead the practitioners into morally ambiguous areas which are devoid of legal determinacy and hence raise tricky moral questions, if they may be called dilemmas.
3. **Loss of Public Trust:** In the meantime, legal loopholes are creating conditions that allow controversy concerning biotechnology practices to continue on ad infinitum and without a clear resolution or public trust—as well as for people who have misgivings about issues of safety relative to non-commercial experimentation.
4. **Inequities Across Jurisdictions:** While it is imperative to understand differences between jurisdictions, the key question that should drive increased national effort and collaboration is one of access equity—where individuals could have disparate opportunities based on geography rather than consistent standard-of-care nationally.

8.14.5 THE MORAL OR ETHICAL IMPLICATIONS OF HUMAN CLONING

Reproductive cloning, the creation of a genetically identical organism, likely reawakens strong moral intuitions, yielding its wide discussion among ethicists as well as scientists and in popular rhetoric [36]. The ethical issues raised by reproductive cloning have to do with the moral basis of human identity, individual freedom, and social implications.

8.14.5.1 *IDENTITY AND INDIVIDUALITY*

The main question concerning the morals behind reproductive cloning is whether it will affect the identity and individuality of the person cloned. Cloning, critics say, erases the distinction that is anatomical to humans. The cloned person would not be seen as a new individual but just another copy or an "add-on" of the original. This idea might create social discrimination where clones are considered to be worthless or lesser than natural-born people. Indeed, philosophers like Paul Ramsey argued that cloning would turn children into projects rather than gifts, as they would have been created to satisfy the desires or needs of someone else. The notion that life can be purchased raises serious ethical concerns regarding cloning and whether such children are the result of genuine maternal love or selfish desires to control or possess another.

8.14.5.2 *AUTONOMY CONCERNS*

There are also concerns about the morality of it as well, including questions around autonomy. Reproductive cloning could place predisposed demands on the clone with regard to their progenitor. If a deceased sibling is cloned, for example, it may have an inbuilt pressure to fulfill the characteristics or accomplishments of the dead child as well. This demand could hinder the clone from creating and following their own unique sense of identity across time. Autonomy, in the most basic ethical theory, is an important principle that each person has the human dignity of making choices in their own life. Cloning would potentially deprive this right as it brings to life an entity whose inheritance is on the essence of a being. This concern is at the heart of the moral philosophy of Immanuel Kant, who said that people should never be treated only as vehicles to an end but always and everywhere must serve as ends in themselves. One risks breaking this ethical maxim by creating a clone because they wanted something.

8.14.5.3 *PSYCHOLOGICAL IMPACTS*

Clones have their own psychological ramifications as well. Besides, it could be mentally very confusing for cloned individuals to figure out who they are. They might wonder: are they even really just a copy of someone else and not their own thoughts and dreams? The psychological burden of these struggles can manifest as issues such as low self-esteem and identity crises.

These eeriness's are only compounded when we consider "genetic determinism" as a concept. Genetics partially determines temperament and character, but contrary to the common belief of many folks, genetics alone does not dictate an individual's personality/behavior. These beliefs create space for further feelings of insufficiency in clones who can feel as if they are living under the shadow of their progenitor; a legacy that looms ominously large. While the clones may have different identities because of what they went through and where they ended up, society will then impede those newly developed individual thoughts [37].

8.14.5.4 POTENTIAL FOR EXPLOITATION

A second major ethical concern of reproductive cloning is the possibility of exploitation of cloned persons and society as a whole. Cloning is placed somewhere along a wide spectrum of motivations, from clones who simply desire companionship and medical advancements to those whose intentions are more malign, driven by selfishness or control.

8.14.5.5 COMMODIFICATION OF HUMAN LIFE

Reproductive cloning has the potential to turn human beings into commodities, rather than gifts created in response to certain needs or wants. That commodification can take many forms, like:

1. **Organ Donation**: A potentially troublesome use of reproductive cloning would be to make clones only for the purpose of having their organs harvested and transplanted. This poses the deeply troubling question of whether it is ethical to have a child for just health reasons.
2. **Replacement Children:** When parents lose a child, they may be tempted to create a cloned version as their "replacement." This situation can create not only an emotional problem, but it is also very likely that the cloned individual may be subject to unrealistic expectations about having a gone due to a deceased child.
3. **Cloning for Human Reproduction and Cloning-for-Biomedical-Research:** Furthermore, the potential overlap between human reproductive cloning in combination with genetic engineering heightens these concerns. For example, if parents were able to

choose features for their cloned offspring (like intelligence or looks), they might treat these people as objects created to fulfill certain criteria rather than accepting them as original individuals.

8.14.5.6 SOCIETAL IMPLICATIONS

The societal concerns of reproductive cloning go beyond individual instances to larger questions about the value society places on human life and identity:

1. **Discrimination:** Clones might be discriminated against due to their "unnatural" origins, causing them to be shunned from society. From employment to social relationships, this discrimination can limit the sense of fulfillment that underscores a life well-lived.
2. **Cloning:** In multiple jurisdictions, the legal status and rights of individuals created by cloning are uncertain. This leads to the question of equal protection under law and possible discrimination on a genetic basis, compared with those who are naturally conceived.

However, society must also bear the ethical responsibility of treating cloned individuals with respect and dignity while creating an environment in which they can live free from burden—particularly emotional burden—from their genetic origins.

8.15 CLONED IDENTITY AND RIGHTS

Another major second-order concern in the debate over therapeutic cloning is that of cloned individuals and their rights/issues. Advances in cloning technologies have raised complex questions about how the legal status of these individuals should be dealt with, from a societal standpoint.

8.15.1 LEGAL STATUS

1. **No Clear Legal Frameworks:** Cloned persons have no status at law in most countries, as human cloning remains largely unregulated. This uncertainty leads to calls for whether existing laws would afford clones equal protections with naturally conceived individuals.

2. **Human:** Legal constructs in many states do not discriminate between persons based on genetic origin; however, clones might find themselves exposed to specific obstacles with regard both to their citizenship and individual rights. Why should they be subject to the same rights as anyone else? Will their biological ancestry become a basis for discriminatory exclusion from certain legal safeguards?
3. **Risk of Discrimination:** Clones will likely be stigmatized or discriminated against for being clones. They could then be seen as not fully human and stigmatized or left out of social circles and opportunities in society.
4. **Jus Soli:** Also, what legal rights stemming from parentage do (or should) apply to the cloned child growing up? These issues illuminate the necessity of a system to regulate parental responsibilities in cases where cloned offspring are involved.

8.15.2 IDENTITY CONCERNS

1. **Confusion Over Identity:** Cloning oneself could send you down an identity crisis, regardless of if the aforementioned person is still living or otherwise. This view of us being more a copy than an original person leads to the self in us feeling less important, and we begin suffering from identity crises as well.
2. **Social Expectations:** Clones could be placed under societal expectations due to the accomplishments or characteristics of their genetic predecessor, restricting them and stifling their independence.
3. **Self-Worth:** Would cloned individuals struggle with their self-worth and what would make them unique? They might think about the fact that they were made scientifically, which could make them wonder where in this world they even belong, and if they are only of value because some person likes their genes.
4. **Psychological Effects:** These are dire because clones will undergo mental stress that can result from the burden of living up to outsider expectations, or they could suffer with the feeling of not being an individual but only a genetic copy without life/positive experiences [38].

8.16 CONCLUSION

The social concerns of therapeutic cloning are not just about rights and identity; they are also linked to questions of equity, public relations, and cultural resistance to biotechnology. Clones are going to be discriminated against simply because society does not understand how cloning works. Perceptions of clones as "unnatural" or "inferior" stem from misunderstandings about what cloning is. This can manifest in discriminatory practices related to both education and employment, which can further filter down into social life.

The future of cloning and genetic manipulation needs to be trodden carefully, with one foot in innovation but another foot rooted firmly in ethics. While we dive into the promising applications of these technologies, from medical breakthroughs to environmental preservation, it is essential that our eyes remain wide open about their ethical consequences. Through developing holistic ethical frameworks rooted in universal dialogue, committing to broad public engagement through educational campaigns, reinforcing legal defenses, and fostering an attitude of equity-in-access, we can assure innovation opportunities by promoting diverse, talented collaborations. This can chart a new horizon that upholds human dignity alongside scientific progress. In the face of these multifaceted problems, it is our promise to ethical treatment that will ultimately decide how we build upon biotechnological breakthroughs for all mankind.

KEYWORDS

- **biotechnology**
- **cloning**
- **CRISPR technology**
- **gene editing**
- **genetic manipulation**
- **human embryo**
- **human reproduction**
- **therapeutic cloning**

REFERENCES

1. Topol, E. J. (2019). High-performance medicine: The convergence of human and artificial intelligence. *Nature Medicine, 25*(1), 44–56. https://doi.org/10.1038/s41591-018-0300-7 (accessed on 5th January 2026).
2. Jiang, F., Jiang, Y., Zhi, H., Dong, Y., Li, H., Ma, S., et al. (2017). Artificial intelligence in healthcare: Past, present and future. *Stroke and Vascular Neurology, 2*(4), 230–243. https://doi.org/10.1136/svn-2017-000101 (accessed on 5th January 2026).
3. Davenport, T., & Kalakota, R. (2019). The potential for artificial intelligence in healthcare. *Future Healthcare Journal, 6*(2), 94–98. https://doi.org/10.7861/futurehosp.6-2-94 (accessed on 5th January 2026).
4. Panch, T., Mattie, H., & Celi, L. A. (2019). The "inconvenient truth" about AI in healthcare. *NPJ Digital Medicine, 2*, 77. https://doi.org/10.1038/s41746-019-0155-4 (accessed on 5th January 2026).
5. Asan, O., Bayrak, A. E., & Choudhury, A. (2020). Artificial intelligence and human trust in healthcare: Focus on clinicians. *Journal of Medical Internet Research, 22*(6), e15154. https://doi.org/10.2196/15154 (accessed on 5th January 2026).
6. Char, D. S., Shah, N. H., & Magnus, D. (2018). Implementing machine learning in health care–Addressing ethical challenges. *The New England Journal of Medicine, 378*(11), 981–983. https://doi.org/10.1056/NEJMp1714229 (accessed on 5th January 2026).
7. Vayena, E., Blasimme, A., & Cohen, I. G. (2018). Machine learning in medicine: Addressing ethical challenges. *PLoS Medicine, 15*(11), e1002689. https://doi.org/10.1371/journal.pmed.1002689 (accessed on 5th January 2026).
8. Nebeker, C., Torous, J., & Bartlett Ellis, R. J. (2019). Building the case for actionable ethics frameworks in digital health. *BMC Medicine, 17*(1), 91. https://doi.org/10.1186/s12916-019-1340-3 (accessed on 5th January 2026).
9. Wachter, R. M., & Cassel, C. K. (2020). Sharing health data and biospecimens with industry – A principle-based approach. *JAMA, 324*(4), 337–338. https://doi.org/10.1001/jama.2020.9883 (accessed on 5th January 2026).
10. Nebeker, C., Harlow, J., Espinoza Giacinto, R., Orozco-Linares, R., Bloss, C. S., & Weibel, N. (2017). Ethical and regulatory challenges of research using pervasive sensing and other emerging technologies: IRB perspectives. *AJOB Empirical Bioethics, 8*(4), 266–276. https://doi.org/10.1080/23294515.2017.1403980 (accessed on 5th January 2026).
11. Mittelstadt, B. (2019). Principles alone cannot guarantee ethical AI. *Nature Machine Intelligence, 1*(11), 501–507. https://doi.org/10.1038/s42256-019-0114-4 (accessed on 5th January 2026).
12. Bender, E. (2021). Dangers of deceptive AI and how to avoid them. *Nature, 592*(7855), S4. https://doi.org/10.1038/d41586-021-00787-5 (accessed on 5th January 2026).
13. Haibe-Kains, B., Adam, G. A., Hosny, A., Khodakarami, F., Waldron, L., Wang, B., et al. (2020). Transparency and reproducibility in artificial intelligence. *Nature, 586*(7829), E14–E16. https://doi.org/10.1038/s41586-020-2766-y (accessed on 5th January 2026).

14. Girijakumari Sreekantan Nair, S., & Chandrasekaran, B. (2024). MetaHealth: Unlocking metaverse technologies in digital healthcare. *Expert Review of Medical Devices*, 1–3.
15. Lakshmi, D., Kondurkar, I., Kumar, R., & Banerjee, R. (2024). Intelligent healthcare systems in the metaverse: Architecture, applications, challenges, and opportunities. In *The Metaverse for the Healthcare Industry* (pp. 17–32).
16. Wang, Y., Zhu, M., Chen, X., Liu, R., Ge, J., Song, Y., & Yu, G. (2024). The application of metaverse in healthcare. *Frontiers in Public Health, 12*, 1420367.
17. Sutradhar, K., Venkatesh, R., & Venkatesh, P. (2024). Smart healthcare services employing quantum Internet of Things on metaverse. In *Healthcare Services in the Metaverse* (pp. 170–189). CRC Press.
18. Nguyen, H. S., & Voznak, M. (2024). A bibliometric analysis of technology in digital health: Exploring health metaverse and visualizing emerging healthcare management trends. *IEEE Access*.
19. Zhang, S., Huang, Y., & Yan, H. (2024). The role of artificial intelligence in healthcare: Opportunities and challenges for the metaverse. *Journal of Medical Internet Research, 26*(2).
20. Patel, V., Dhillon, M., & Ramesh, A. (2024). Enhancing patient care through virtual reality in hospitals. *Journal of Hospital Medicine, 19*(1), 12–19. Allen, M., & Gomez, D. (2024). Virtual reality as a tool for pain management in hospitals: A metaverse perspective. *Pain Management, 14*(2), 91–98.
21. Peterson, L., & Davis, K. (2024). Ethical frameworks for patient consent in virtual environments. *Journal of Law, Medicine & Ethics, 52*(1), 24–30.
22. Nguyen, T., & Pham, Q. (2024). Leveraging the metaverse for remote healthcare: Opportunities and risks. *Telemedicine and e-Health, 30*(2), 129–137.
23. Yang, F., & Chen, X. (2024). The impact of the metaverse on hospital service quality: A mixed-methods study. *BMC Health Services Research, 24*(1), 45–52.
24. Singh, B., & Kaunert, C. (2024). Adventure in high altitude of mountainous topographies and health impacts: Lensing tourism sustainability via reducing ecological and sociocultural footprint and health emergency and medical assistance management. In G. Meraj, S. Hashimoto, & P. Kumar (Eds.), *Navigating Natural Hazards in Mountainous Topographies* (p. 15). Cham: Springer. https://doi.org/10.1007/978-3-031-65862-4_15 (accessed on 5th January 2026).
25. Singh, N., Lal, S., Raghav, A., Arora, M. K., Singh, B., & Kaunert, C. (2024). Power of ML and DL in precision modeling and climate predictions: NextGen ocean insights. In B. Riswandi, B. Singh, C. Kaunert, & K. Vig (Eds.), *AI Applications for Clean Energy and Sustainability* (pp. 254–273). IGI Global. https://doi.org/10.4018/979-8-3693-6567-0.ch013 (accessed on 5th January 2026).
26. Singh, B., Kaunert, C., Lal, S., Arora, M. K., & Singh, G. (2024). Abundance ocean wave energy to electricity with artificial intelligence and IoT solutions: Future of clean energy and sustainability in smart cities. In B. Riswandi, B. Singh, C. Kaunert, & K. Vig (Eds.), *AI Applications for Clean Energy and Sustainability* (pp. 274–298). IGI Global. https://doi.org/10.4018/979-8-3693-6567-0.ch014 (accessed on 5th January 2026).
27. Lal, S., Arora, M. K., Riswandi, B. A., Singh, N., & Singh, B. (2024). Melding artificial intelligence and remote sensing data for real-time monitoring of marine biodiversity.

In B. Riswandi, B. Singh, C. Kaunert, & K. Vig (Eds.), *AI Applications for Clean Energy and Sustainability* (pp. 299–319). IGI Global. https://doi.org/10.4018/979-8-3693-6567-0.ch015 (accessed on 5th January 2026).

28. Arora, M. K., Lal, S., Singh, B., & Olyaee, M. (2024). Through an energy lens: Examining SDGs from a policy perspective. In B. Riswandi, B. Singh, C. Kaunert, & K. Vig (Eds.), *AI Applications for Clean Energy and Sustainability* (pp. 320–343). IGI Global. https://doi.org/10.4018/979-8-3693-6567-0.ch016 (accessed on 5th January 2026).
29. Kaushik, T. K., Ravish, S., Singh, A., Raghav, A., & Singh, B. (2024). Potential of artificial intelligence in the agricultural industry: A comprehensive review. In B. Riswandi, B. Singh, C. Kaunert, & K. Vig (Eds.), *AI Applications for Clean Energy and Sustainability* (pp. 344–365). IGI Global. https://doi.org/10.4018/979-8-3693-6567-0.ch017 (accessed on 5th January 2026).
30. Raghav, A., Singh, B., & Jermsittiparsert, K. (2024). Role of artificial intelligence and machine learning in clean and blue energy for enhancing efficiency and sustainability. In B. Riswandi, B. Singh, C. Kaunert, & K. Vig (Eds.), *AI Applications for Clean Energy and Sustainability* (pp. 366–386). IGI Global. https://doi.org/10.4018/979-8-3693-6567-0.ch018 (accessed on 5th January 2026).
31. Robinson, T., & Williams, A. (2024). The metaverse in medical education: Reshaping the future of learning. *Medical Education, 58*(4), 367–375.
32. Wang, M., & Zhang, H. (2024). Privacy-preserving technologies in the metaverse: Implications for healthcare. *IEEE Journal of Biomedical and Health Informatics, 28*(3), 560–568.
33. Ali, S., & Khan, J. (2024). The promise and peril of virtual reality in patient care. *Journal of Patient Experience, 11*, 201–208.
34. Brown, L., & Harris, P. (2024). The metaverse in clinical practice: Enhancing diagnosis and treatment. *American Journal of Medicine, 137*(1), 23–30.
35. Garcia, M., & Rodriguez, L. (2024). Ethical dilemmas in metaverse-based healthcare: Balancing innovation and patient rights. *Journal of Medical Ethics, 50*(3), 167–174.
36. Thompson, G., & White, R. (2024). Scaling metaverse solutions for public health: Lessons from early adopters. *Public Health Reviews, 45*(2), 205–213.
37. Zhao, R., & Tang, J. (2024). Blockchain in the metaverse: Enhancing trust in digital healthcare. *Journal of Medical Systems, 48*(1), 12–21.
38. Wong, B. K., Vengusamy, S., & Chua, C. S. (2024). Digital transformation of healthcare services and infrastructure in Asia: The metaverse bound? In *Digital Healthcare in Asia and Gulf Region for Healthy Aging and More Inclusive Societies* (pp. 283–305). Academic Press.

CHAPTER 9

Genomic Innovations in Drug Discovery and Clinical Medicine

SHAYUT PAVAPANUNKUL[1*] and ANJALI RAGHAV[2*]

[1]*Universitas Muhammadiyah Sidenreng Rappang, Sidenreng Rappang, South Sulawesi, Indonesia*

[2]*Sharda School of Law, Sharda University, Greater Noida, Uttar Pradesh, India*

*Corresponding authors

ABSTRACT

These genomic breakthroughs set off the paradigm shift in drug discovery and clinical medicine, providing clinicians with diagnostic tools unavailable heretofore for understanding the molecular basis of disease. Genomics has revolutionized molecular biology, and Brookhaven National Laboratory is crucial in supporting foundational research. Still, its downstream implications have put science at the center of modern discovery-based healthcare. Technological advancements in the field of genomics, such as next-generation sequencing (NGS), genome-wide association studies (GWAS), and CRISPR, coupled with subsequent identification of genetic drivers for diseases, have paved the way for precise therapeutic interventions to treat a wide range of other clinical conditions. Genomic databases, bioinformatics, and artificial intelligence (AI) have made drug discovery more efficient, identifying potential drug molecules in less time. This also speculates about the future, covering trends such as multi-omics integration, gene editing breakthroughs, and AI-driven predictive models that could underpin further advances. Genomic medicine has the potential to truly revolutionize both patient care and drug development by enabling

Genetic Editing for Healthcare: Advances in Disease Management.
Bhupinder Singh, Rishabha Malviya, Sahil Lal, and Sathvik Belagodu Sridhar (Eds.)

DOI: 10.1201/9781779643827-9

more effective, targeted, personalized healthcare solutions via pioneering research efforts coming from across the globe.

9.1 INTRODUCTION

The field of genomics, the study that discusses all DNA in an organism, the genome, has been fundamental to modern biomedical research. Genomic science, the analysis of an individual's genetic information encompassing all genes and elements found in a cell, has provided billions to trillions of unprecedented insights into diseases over the past few decades and has remodeled several aspects of healthcare. Genomics can tell us what the sequence of our DNA is, how genes work, and give insights as to which genetic changes are associated with specific health conditions [1]. This level of gene specificity has paved the way for significant successes in drug discovery and clinical practice. Drug discovery is finding new medications that may act as effective compounds for treating certain diseases by inhibiting disease pathways. Traditionally, this has been done through a trial-and-error process with high failure rates that took months to years. Including genomic data has changed the viewpoint from which one looks at discovering drugs. This allows researchers to understand the genetic pathways of diseases better and find those crucial drug targets much faster. However, it has also helped in creating refined and specific drugs for a particular population of patients. The infusion of genomics into clinical medicine has yielded a new edge in personalized or precision medicine. Customized health care, personalized medicine, uses genetic information for preventive measures, diagnosis, and treatment. Under the umbrella of personalized healthcare, customized medical care based on a patient's unique genetic code has enabled us to treat more effectively and minimize drug side effects. This means that physicians can determine how a specific therapy will likely work and respond in similar patients, resulting in more efficient, safe medical interventions [2].

This chapter investigates the catalytic effect of genomics on drug discovery and clinical medicine, illustrating significant innovations that have surfaced. In the first part of this chapter, we will address how essential and influential genomic technology, especially next-generation sequencing (NGS) and genome-wide association studies (GWAS), contributed to accelerating drug discovery since it helped with the identification of novel pharmacological targets as well as understanding more deeply pathophysiologic background for diseases [3]. One of the most significant benefits of personalized medicine is that it offers a pharmacogenomic approach to treating prescribed drugs, allowing for more effective therapeutic outcomes and reducing adverse drug reactions. Though benefits are many, ethical issues alone could put a Cardiac 3D Print on the same footing as Ambergris (whale vomit), endangered species curing voodoo snake oil, or illicit drug skunk, which costs billions to make legal, let alone in dollars spent subsidizing avenues around it. The people who work at NIH are front and center; they will touch on cases where drug development has succeeded by leveraging knowledge obtained through genomics, including a deep dive into the journey from gene identification to drug FDA approval. The second half will be about the impact of genomics in clinical medicine, especially personalized medicine. The discussion will focus on how genomic data is increasingly being used to personalize treatment decisions and, in doing so, enhance both effectiveness and safety. Individual case studies incorporating pharmacogenomics into standard medical practice to tailor drug prescriptions will also be presented [4]. The chapter will also discuss new cutting–edge technological developments in *CRISPR* gene editing and artificial intelligence that are significant schema shifts reshaping drug discovery and genomic medicine to using genomic data for drug discovery and clinical practice. This covers data privacy, the high cost of genomic technologies, and the regulatory environment. This chapter intends to analyze these challenges that must be addressed in the realm of genomic innovations and try to provide a holistic view of where genomics is today vis-à-vis healthcare, thereby paving the way for what we can expect from genomic applications in the future. Genomics has been the most valuable resource in modern medicine, revolutionizing drug discovery and fostering personalized treatment explicitly designed to meet each individual's genetic composition [5]. This chapter examines these advances and shows the future where genomic innovations are paramount to shaping healthcare.

9.2 SIGNIFICANCE OF GENOMICS

In drug discovery and clinical management, genomics has dramatically changed how diseases are perceived and cured. In the paradigm of this revolution lies the potential to recognize and interpret information about genetic variations propelling disease. The insights garnered have been critical to discovering and delivering new medicines that modulate disease by addressing their root cause in various common cancers, cardiovascular diseases, or rare genetic diseases. In cancer, for example, genomics has led to the development of personalized treatments that specifically target mutations in genes found only in tumors but not normal cells and have thus reduced adverse side effects [6].

The discovery of biomarker molecules that reflect the presence, absence, or severity of a disease state by genomics is one of its most valuable contributions to drug development. Researchers can create drugs for specific targets using genetic markers linked to certain conditions. This has, in turn, given rise to the development of target-specific therapies aimed at interfering directly with disease mechanisms for better efficacy and lower toxicity. Examples of this are the development of treatments for HER2-positive breast cancer and cystic fibrosis, where drugs target specific mutations identified in patients. Also, genomics has bred a whole new pharma kitty to think about pharmacogenomics, the study of how an individual's response to medication varies because of his or her genetic inheritance [7]. Genetic variants that affect the metabolism and response to medication help explain why the therapeutic use of drugs varies among patients. Understanding these genetic differences can help clinicians prescribe medications that offer the most potential for treatment with reduced adverse drug reactions. This has been significant in oncology, where genetic testing can now predict the best chemo for a patient or is increasingly utilized to guide antidepressant selection in psychiatry [8].

9.3 ROLE OF GENOMICS IN DRUG DISCOVERY

Genomics has dramatically impacted drug discovery by providing tools for gaining insight into the genetic and molecular bases that lead to many diseases. Among the most valuable uses of the genome is to gain insights into which individual genes, variations, and pathways contribute to or cause several diseases. This has been critical for pinpointing possible drug targets that can be used to develop more selective and potent therapeutic compounds. The identification of disease pathways is one fundamental way that genomics helps in the process of drug discovery [9]. Genomic research makes this possible, allowing scientists to follow the detailed pathways of genes and proteins involved in diseases. This enables researchers to pinpoint central nodes within the path disrupted in disease states, offering potential drug intervention sites. This allows therapies to be developed that can interrupt disease pathways at a molecular level, so these treatments work in the kindest possible fashion. This is where DNA sequencing comes into play [10]. The advancement of NGS has enabled whole genome sequencing to be performed rapidly at a relatively low cost. Since then, the genomes of patients with various diseases have become available for researchers to examine the genetic variation that may underpin susceptibility. The sequencing of cancer genomes, for example, has identified many mutations that can fuel tumor growth and thus have a therapeutic value in targeted therapies to more directly inhibit these mutated products. GWAS and Drug Development This is a screenshot of the email after performing suitable fine mapping in large-scale GWAS, that is, scanning genomes of thousands of individuals to find genetic markers related to certain diseases. These identified genes are putative targets for drug development. GWAS has been utilized to identify genes associated with autoimmune diseases (such as rheumatoid arthritis and Crohn's disease), leading to the development of drugs that target immune pathways involved in these conditions. CRISPR-Cas9, a powerful gene editing technology developed by Harvard geneticist George Church, could contribute to unlocking new pathways in drug discovery as its ability allows scientists to modify genes accurately [11]. It can be used to create disease models by integrating specific genetic mutations into cells or organisms to study mutant phenotypes and explore the mechanism of diseases. CRISPR has made it possible to correct disease-causing mutations, which will be necessary for developing gene therapies. Among the applications of these CRISPR technologies is their use in creating therapies for genetic disorders such as sickle cell anemia, where the technology may be used to correct a mutation that causes the disease [12].

9.4 HIGH-THROUGHPUT SCREENING (HTS) AND GENOMIC DATABASE

High-throughput screening (HTS) and genomic databases have greatly facilitated the integration of genomics into drug discovery. HTS enables researchers to rapidly test thousands of compounds against a target and identify those with the potential for therapeutic use. When paired with genomic information, HTS could be applied to identify molecules that would act on targets recognized by genomics and thus help make the drug discovery effort more rapid. Since then, however, genomic databases have evolved into a staple tool researchers utilize to pinpoint compounds with drug-compounding potential. This data encompasses thousands of databases with enough genetic information, DNA sequences, and gene expression data, which are known to name just a few areas, especially regarding variant-related biology. Researchers may then use this data to look for overexpressed or mutated genes in disease states, which are drug target candidates [13]. The Cancer Genome Atlas (TCGA) provides complete genomic profiles of multiple cancer types, making it possible for scientists to discover several genes altered in tumors relatively frequently. Therefore, this may provide novel targets for drugs to target these genes specific to the working of cancer. Computational biology and bioinformatics have also made significant progress in the past few decades, improving their processes to analyze larger genomic datasets. Machine learning algorithms will start to be used to find patterns in genomic data that might reveal disease mechanisms. This has allowed the discovery of novel drug targets that might have gone underappreciated had they been found through traditional means [14]. Using bioinformatics tools, we can predict how a drug will likely interact with different genetic targets, resulting in better drug design and less time and cost getting new ones to market. In particular, the holism of this data means that integration with other types of biological big data (such as proteomics and metabolomics) leads to a more complete orientation toward disease biology. This integrated multi-omics approach identified multiple biomarkers predicting disease progression or response to therapy. One way to both directly study how genetic mutations result in changes in protein expression by targeting those specific proteins with drugs is to integrate genomic and proteomic information [16].

9.5 VARIOUS CASE STUDIES

Genomic innovations are evident in numerous therapeutic areas regarding how drug discovery can be reformed. The following case studies show how genomic information has been used to create targeted cancer therapies, develop drugs for various genetic conditions, and much more are:

1. **Cancer: Trastuzumab (Herceptin):** It is a monoclonal antibody developed for breast cancer patients with tumors overexpressing the HER2 protein. This is because genomic research has shown that approximately 20% of all breast cancer patients have their disease driven by amplification of the HER2 gene, resulting in increased levels of the tumor-activating protein product known as

HER2. Trastuzumab binds to HER2 protein on the surface of breast cancer cells and blocks their activity, developing a stopover in tumor growth. By targeting the root of cancer at a genetic level, this tailored therapy has greatly enhanced patient outcomes for HER2+ breast cancers and is an excellent example of how genomics can help develop individualized biological treatments [17].

2. **Cystic Fibrosis: Ivacaftor (Kalydeco):** CF is an inheritable condition, and there are multiple gene mutations responsible for creating a protein in cell membranes that control salt passage. Identifying the set of genetic mutations that result in CF has since led to the drug ivacaftor, which targets one mutation (G551D) within the *CFTR* gene. Ivacaftor increases the activity of this defunct CFTR protein, resulting in improved lung function and quality of life for patients with that specific mutation. This case illustrates that if the genetic basis of a disease is well understood, treatments can be developed quickly to target particular mutations [18].
3. **Cardiovascular Disease: Pcsk9 Inhibitors (Alirocumab, Evoldcumar):** Genetic studies have pointed to the critical role of PCKS9 in regulating cholesterol. People carrying mutations that impair PCSK9 function have low cholesterol levels and a low risk of heart disease. Now, PCSK9 inhibitors like alirocumab and evolocumab, or human monoclonal antibodies, block the protein action of this gene, resulting in controlled LDL levels. This is a powerful example of how genomic discoveries can translate into effective new treatments, giving patients unable to reach their cholesterol goals on traditional statins another medication option [19].
4. **Gene Therapy: A Hope for Rare Genetic Disorders:** Spinal muscular atrophy (SMA) is a genetic neuromusculoskeletal disorder caused by mutations in the *SMN1* gene, which results in loss of motor neurons and muscle weakness. The genetic foundation of SMA was determined by genomic studies, which prepared the way for creating gene therapies, including onasemnogene abeparvovec (Zolgensma). The treatment supplies patients with a functional form of the *SMN1* gene, allowing their bodies to manufacture the deficient protein and ease motor symptoms. This shows the potential of genomics in curing genetic conditions once thought to be untreatable [20].

9.6 GENOMES TO BEDSIDE PERSONALIZED MEDICINE

Personalized medicine is a medical model that proposes customizing healthcare, with medical decisions tailored to individual patient characteristics. Personalized medicine moves away from the "one-size-fits-all" approach, which says that everyone with a particular disease will respond similarly to one or more safe and effective treatments for this condition while ignoring significant genetic, environmental, and lifestyle differences between patients. This became a guiding principle in present-day medicine, mainly when genomics made deciphering the interplay of genes and health or disease easier. Decreased response variability in clinical trials: The value of personalized medicine is in reducing patient variance, thereby enhancing the predictability and efficacy of treatments. Knowing an individual's genetic information allows clinicians to tailor treatments that are likely to work (hopefully better than trial-and-error medicine) with fewer side effects. Personalized medicine greatly facilitates preventive healthcare, allowing prevention measures in populations with high risk for certain diseases and induction of early intervention strategies [21].

At the foundation of precision medicine is genomics. The human genome is just the entirety of our genetic information that tells how a cell in your body functions. Genomics allows us to understand how different drugs can be metabolized in various ways by individuals since the genome varies between us and is essential for precision medicine. Such differences can occur as genetic polymorphism (e.g., they influence how our bodies handle and affect drug absorption, distribution, metabolism, or elimination) changes how drugs interact with therapeutic targets from one individual to another. Using an individual's genetic makeup, healthcare providers can tailor drug and dose selection to reduce adverse effects while maximizing therapeutic benefit [22]. NGS technology has helped transform personalized medicine by enabling a patient's rapid and relatively inexpensive genome-wide sequencing. This has allowed researchers to discover genetic differences that can raise or lower the risk of getting certain diseases or change how people react to a drug. In addition, large-scale genomic projects like the Human Genome Project have produced rich databases for clinicians to integrate specific genetic information with disease susceptibility and drug response to enable precision medicine.

9.7 PHARMACOGENOMICS

Pharmacogenomics analyzes how genes affect a person's response to medicine. This exciting field mirrors the interface between genomics and pharmacology, where drug therapy is tailored to a patient's genetic code. The ultimate objective of pharmacogenomics is to ensure that the perfect medication can automatically be used for each patient and that it is a safe one custom-made according to the genetic makeup, consequently sparing the public from prolonged therapy before surgical prescribing. Genomics is essential in understanding pharmacogenomic drugs, ranging from drug efficacies to adverse reaction studies. These genetic variations, such as single nucleotide polymorphisms (SNPs), can result in changes to metabolism by drug-metabolizing enzymes, target binding, and transporter uptake of drugs. For example, differences in the CYP450 family of enzymes that metabolize many drugs may account for different rates at which individuals can process these molecules. Some genetic variants in patients might cause the rapid metabolism of drugs, which means they do not work as well, or their slow metabolism, which leads to toxicity [23]. Doctors can use these genetic differences to modify drug dosage or even choose different drugs to produce an optimal therapeutic result. Also, pharmacogenomics can predict what type of adverse drug reactions will occur as a by-product. Side effects (adverse drug reactions) cause a considerable degree of morbidity and mortality, as well as extra healthcare costs to manage these side effects. The knowledge of the genetic contributions to these reactions allows healthcare providers to identify at-risk patients and intervene proactively [24]. For example, individuals with specific genetic characteristics may suffer more severe side effects from the use of chemotherapy drugs. If clinicians can identify such patients via genetic analysis, they can tailor the treatment plan accordingly and mitigate those risks.

9.7.1 IMPACT IN CANCER BY PHARMACOGENOMICS

Targeted therapies, as the name suggests, target specific genetic mutations that drive cancer cell growth without harming normal, healthy cells. In addition to improving treatment efficacy, this strategy may also reduce toxicity. For example, drugs like imatinib (Gleevec) have been used to directly target the specific genetic mutations observed in certain leukemias, translating into better patient outcomes [25].

9.8 IMPACT OF GENOMICS ON PATIENT CARE

Several clinical cases have exemplified the impact of genomics on patient care in real-world application. Personalized therapy examples are as follows:

1. **HER2-Positive Breast Cancer:** One of the most celebrated success stories in personalized medicine is trastuzumab (herceptin), which was developed to treat HER2-positive breast cancer. HER2 is a protein that makes breast cells grow, and in one out of five cases, the *HER2* gene causes the overgrowth of cancer tumors. By sequencing the genomes of tumors, clinicians can identify patients with overexpression of HER2 and use trastuzumab, a drug engineered to target this particular protein. One example is the successful targeting of HER2-positive breast cancer with historically high rates of lung metastasis by developing medicines that reflect genetic insights into tumors [26].
2. **Cystic Fibrosis (CF):** A genetic disorder caused by a mutation in the CFTR gene, which encodes a protein responsible for regulating salt and water movement between cells. Genetic advances have enabled the development of drugs that target individual mutations to correct defects at many levels, from cellular trafficking and membrane insertion through channel gating. An example is ivacaftor (Kalydeco), a drug that potentiates the function of the defective protein in patients with G551D mutation. Such a treatment, tailored specifically to this mutation of CF, has proven unprecedented improvements in lung health and quality of life among these patients. Furthermore, the creation of combination therapies like lumacaftor and ivacaftor (Orkambi) has opened up treatments for those with other mutations, demonstrating how personalized medicine can better outcomes even in genetic diseases [27].
3. **Warfarin Dosing:** Warfarin is a commonly prescribed medication for reducing the risk of blood clots, yet effective dosing varies significantly from person to person. Serious complications, such as bleeding or clot formation, can occur if incorrect dosages occur. Pharmacogenomic testing has identified variants in the *VKORC1* and *CYP2C9* genes that influence warfarin metabolism. Clinicians can use pharmacogenetic tests to identify these genetic

variants before initiating treatment and then adjust the warfarin dose accordingly in each patient, minimizing adverse events while increasing the therapeutic efficacy of this medication [28].

4. **Melanoma: BRAF Inhibitors:** Genomic investigation has unveiled that up to 50% of patients diagnosed with melanomas carry mutations in the BRAF gene. These mutations cause the BRAF protein to be permanently activated, which stimulates cancer cell growth. Scientists have developed vemurafenib (Zelboraf) and dabrafenib (Tafinlar), which target the mutant BRAF protein, slowing down tumor growth. As a result, patients with melanoma are routinely screened for BRAF mutations and, when they test positive, can be treated with these targeted therapies that have demonstrated better survival than traditional chemotherapy [29].
5. **Precision Oncology: Tumor Profiling:** The evolution of genomics has made it possible to sequence the entire tumor genome for a comprehensive profile that shows what mutations exist in any given cancer and which are driving its progression. This knowledge helps oncologists deliver precision cancer care; they can offer multiple therapies approved for treating other cancers that might specifically target the mutations in a patient's tumor. The ability to provide individualized treatment for a specific type of lung cancer, certain types of gastrointestinal stromal tumors (GISTs), and some other cancers that are based on the genetic profile of one's tumor progress in more targeted therapy options with improved efficacy and fewer side effects [21].

9.9 GENOMIC MEDICINE IS MAKING STRIDES WITH ADVANCEMENTS IN TECHNOLOGY

9.9.1 NEXT GENERATION SEQUENCING (NGS)

The advent of a new suite of sequencing technologies revolutionized genomic sciences, referred to as NGS. In contrast to traditional Sanger sequencing (which sequences stretches of DNA one base at a time), NGS allows millions of fragments within the same run, massively reducing both expense and processing times required for complete whole-genome sequence data. This has revolutionized genomics research, drug discovery, and clinical diagnostics [22].

NGS technology has dramatically expedited genomics research by making it easier and faster for scientists to sequence genomes. Today, genetic variation can be explored at a population level, and disease-associated mutations for complex traits can be pinpointed with more finesse than ever before. Such deep knowledge has provided new drug targets and treatments. For example, NGS can pinpoint the exact genetic mutations that cause cancers (or other conditions) to formulate targeted drugs that selectively suppress these same mutations [23].

NGS has also been instrumental in drug discovery, enabling the identification of druggable targets by understanding diseases at a genetic level. It means that disease-associated mutations can be quickly identified, allowing researchers to concentrate their resources on developing therapies designed to disrupt these individual genetic changes. Moreover, NGS has enabled the ability to perform massive GWAS designed for finding genetic variants tied with particular traits or conditions. For example, NGS has played a crucial role in developing targeted therapies for cancers harboring specific mutations (e.g., BRCA1/BRCA2 mutations in breast and ovarian cancer). Knowing the mutations driving these cancers means scientists can design treatments that attack those mutated genes, providing more precise and effective treatment [24].

NGS has also transformed clinical diagnostics, allowing us to identify genetic mutations that may be causing inherited diseases or predisposing us to certain conditions. With early warning and intervention, clinicians are better positioned to deliver individualized care determined by a person's genetic makeup. In this manner, NGS is applied in prenatal screening to recognize different hereditary issues, that is, Down disorder, available inside hatchlings. It is also employed to diagnose rare genetic disorders in patients who might have been living for years with elusive symptoms, offering previously inaccessible answers via conventional diagnostic means. Second, because of the ability to rapidly and accurately sequence an individual's complete genome, clinicians can determine mutations that confer drug resistance to know which drugs will work best for a particular patient [30].

9.10 CRISPR (CLUSTERED REGULARLY INTERSPACED SHORT PALINDROMIC REPEATS) AND GENE EDITING

The discovery of CRISPR has fueled a revolution in gene editing and, thus, genomics. The technology lets scientists alter the DNA of living organisms, including humans, with extreme precision. CRISPR is versatile and used in everything from basic research to drug development and gene therapy, making it one of the most promising tools in genetic engineering.

One such impact of CRISPR technology is its effect on drug discovery and development. Meddling with the spread of mutant genes lets them learn more about what genes do and how they sometimes contribute to disease. This has led to the discovery of novel drug targets and the design of precision medicine, which helps target genetic mutations. One use, for example, is the creation of disease models by using CRISPR to introduce specific mutations into their genomes. They mimic human diseases and allow the examination of disease mechanisms and investigational treatments. Similarly, CRISPR was the only method that could be used to screen for genes contributing to drug resistance, and therefore, it might help researchers develop ways around this obstacle. CRISPR has led to gene therapies, which focus on fixing a patient's genes to cure or treat genetic disorders. This is especially life-changing for diseases with a single gene mutation as the cause, such as sickle cell anemia, cystic fibrosis, or muscular dystrophy [31]. These diseases have genetic roots, and scientists can correct defective genes using CRISPR. For example, some researchers use CRISPR in ongoing clinical trials to correct the mutation that causes sickle cell anemia by editing a person's stem cells. Initial results from these trials have been promising, putting patients into remission of disease symptoms.

The uses of gene editing in the clinic extend far beyond rare genetic diseases. The pipeline for CRISPR as a gene-editing tool is wide open, and research indicates that if the advent of flowing genes (such as was observed recently) could be destabilized, this represents many more opportunities to make cancer cells vulnerable to existing therapies or even destroy them directly. It can also be leveraged to engineer immune cells like T cells to spot better and kill cancer, a platform used successfully in clinical trials for some forms of leukemia. With their continued development and innovation, one day, the technology could help with anything from genetic diseases to infectious diseases or cancer [32].

Although highly promising, CRISPR technology is fraught with ethical and technical challenges. Practical issues concerning off-target effects, which means that other parts of the genome are edited along with-like consequences, are questions that point to germline editing risks, meaning potential should also be solved before CRISPR can be widely used in clinics.

However, as tools for directly altering the genome, they still represent an essential step toward making those capabilities available, and research is continuing to try to make CRISPR an even more exact technique [33].

9.11 ARTIFICIAL INTELLIGENCE (AI) AND MACHINE LEARNING (ML) IN GENOMIC MEDICINE

Genomic medicine relies heavily on AI and ML, necessitating vast amounts of genomic data mining. The incorporation of AI and ML to gain further knowledge of our genomics data leads us toward identifying patterns, making predictions, and making decisions using those datasets, accelerating the way drugs are being developed, and helping in finding particular treatments called personalized medicines.

AI and ML are used in genomic data analysis to identify genetic variants that could cause disease risk and mimic or inform how those changes affect protein function and cellular processes. Researchers establish biomarkers associated with various diseases by training machine learning algorithms on big data corpora to create diagnostic tests and companion therapeutic targets. An ML algorithm, for instance, can scrutinize a large quantity of sequencing data to figure out how specific mutations statistically correlate with drug resistance and point researchers toward counteracting strategies [34]. For example, the detection of drug targets based on gene expression data where AI can help identify new therapeutic agents. Their latest data suggests that predictive modeling for drug response and resistance remains one of the incredibly impactful applications of AI and ML in genomic medicine. One can train the predictive models on genomic data, which will help forecast how a specific drug will work with different patients, enabling clinicians to choose an appropriate therapy for each patient. AI models have been employed, for instance, to predict the response of cancer patients with various genotypes (determined from genetic tests) post-administration of chemotherapy and immunotherapy regimens and identify those who can benefit or undergo severe toxicities. This could save clinicians trial-and-error in treatment and improve patient outcomes while reducing healthcare costs. AI has even found its way into drug discovery to help identify new reagents. Conventional drug discovery is a long and expensive process that typically requires screening thousands of compounds to identify active ones for a given target. This is where AI can step in, predicting with high certainty which compounds interact only based on their chemical structure, thereby cutting down costs related to discovering new drugs. Artificial intelligence algorithms create new molecules designed to attach to specific proteins, which could also help develop other medicines (like for Alzheimer's or diabetes) [35].

Also, the fusion of AI with NGS has paved the way for precision medicine platforms that consolidate genomic, clinical, and environmental information for a holistic view of patient health. Many of these platforms can analyze the data and offer insights into which treatment options may best help patients, considering their genetic profile, disease history, and other parameters. Systems like these are already in use for oncology, giving oncologists a way to create treatment plans that are tailored specifically to their patients suffering from cancer, and also for diagnosis of rare genetic disorders where they can support clinicians identifying the root cause behind why an individual is showing symptoms. Thus, NGS and CRISPR technologies combined with AI/ML could revolutionize genomic medicine with the potential for life-saving diagnoses, therapy, and preventative strategies for many diseases. In addition to accelerating the progression of basic science research, these technologies have enabled new therapeutic modalities that previously were unimaginable. Combining these technologies as the field progresses into other areas is expected to achieve even more in genomic medicine, thus benefitting patients worldwide [36].

9.12 INTEGRATING GENOMICS INTO DRUG DISCOVERY

9.12.1 ETHICAL AND REGULATORY ISSUES

Genomics has become increasingly important in drug discovery, contributing to various new and powerful developments while raising ethical and regulatory questions. One of the most critical moral issues concerns preserving privacy and ensuring data security. The extraction of genomic data is highly sensitive, and any breach can result in the misuse of private health information. Unlike other types of medical data, genomic information is about the individual. It cannot just disclose his or her likelihood to develop certain diseases but also top-secret family facts. The nature of genomic data further complicates the issue of consent because it details exactly how a person could come down with diseases to which they may be susceptible. Proper informed consent from individuals should require them to understand what will happen with that shared and protected information. Despite prohibitions such as the Genetic Information Nondiscrimination

Act (GINA) in the U.S., there remain fears about possible future misuse by employers, insurers, or others of a person's genetic data to discriminate against them because of their genes [37].

This has implications for regulatory bodies, such as the U.S. Food and Drug Administration (FDA) or European Medicines Agency (EMA), which must adapt their policies to approve drugs developed around genetic information into clinical care. The challenge is balancing the imperative for strong safety and efficacy testing with new methods of speeding more targeted, personalized treatments down a development pipeline. Regulatory frameworks must also address the heterogeneity in the quality of genomic scale test results, such as whether a patient has been tested for eligibility to receive a particular therapeutic matched at their gene presentation status, and have mandated that any tests used are robust, accurate, and normalized between testing services/laboratories [38]. It is essential to carry out gene-editing technologies, like CRISPR, under proper guidelines to be used ethically in drug development. The regulatory frameworks should also be capable of preventing gene editing applications from being used irresponsibly and/or unethically, with well-defined boundaries on what is permissible.

9.12.2 TECHNICAL AND FINANCIAL CONSTRAINTS

These technical and cost constraints pose a significant barrier to the wide-scale implementation of genomics in drug discovery, as substantiated by the number of testbed-type demonstrations on any genomic approach. The biggest technical hurdle is that genetic data is quite complex. The human genome is massive, and understanding how millions of genes interact with one another or contribute to disease remains daunting. Even though advances have reduced the time and cost needed to sequence entire genomes, it is still difficult to interpret this data. The computational resources and a solid biological background draw the line between "well-pruned" genetic variations and their inverse [39].

Further, genomic-based therapies are resource-intensive to develop. Discovering drug targets, confirming those as "reality," and developing potent agents intervening at the identified loci are all labor- and resource-intensive processes. Standardizing these treatments decreases their availability and, with it, the rationality issue enjoyed by customized drugs. So, a natural technical barrier must be changed without physical test infrastructure for genomic-level integration in clinical practice.

Suppose genomes are to be sequenced in hospitals and clinics. In that case, we need those institutions to be capable of conducting genetic tests, analyzing the results of these analyses on bioinformatics tools or genome browsers, and applying genomic information directly in patient care. Nonetheless, it may not be feasible for many of these tools to be adopted by healthcare facilities in low- and middle-income countries due to a lack of resources or expertise. In addition, healthcare providers must be trained to make sense of genetic information and apply that data to treatment decisions. Genomics has much to offer drug discovery and personalized medicine but will have little impact without the proper infrastructure and education [40].

The other significant problem is the standardization of data. When used and compared, the data drawn from different sources will be effective in genomic research. However, there is no uniform standard for collecting, analyzing, and reporting data, which creates discrepancies and hampers collaborative endeavors. Standardization of the genomic data and protocols is crucial to ease data sharing among research institutions, healthcare providers, pharmaceutical companies, and so forth, and documentation standards are distributed across organizations.

9.12.3 OVERCOMING CHALLENGES

Solving these ethical, regulatory, technical, and financial issues demands a holistic solution. Collaboration between the stakeholders is necessary for research and clinical communities, including regulatory bodies, pharmaceutical companies, and patient advocacy groups, to develop appropriate modalities for implementation that are effective yet ethical. To solve privacy and data security problems, such as the Capital One data breach, we have to work on building robust encryption strategies for our sensitive data. Increased regulation around data sharing and consent: To implement more stringent rules about how patients are informed regarding their data being used, in conjunction with easing standards for privacy breaches so that the tools that come out of these research efforts can be openly used.

Governments and regulatory agencies are trying to overcome these hurdles by establishing more adaptive frameworks that can evolve with the rapid pace of change we witness in genomic medicine. One specific case in point is the adoption of new approval pathways by the FDA to support gene therapies, which present unique challenges. Regulatory agencies are also working more closely with each other internationally to converge standards and practices, which might shorten the drug approval process based on genomics. It is designed to speed development while maintaining the safety and efficacy of new patient therapies. However, more work is needed to make these technologies affordable and available to consumers. The Schwab team added that instead of overhauling the FDA's process altogether, investing more in research and development could help speed up treatments' time to market while creating economies of scale. Here, public-private partnerships can often prove invaluable, providing a mechanism to leverage the resources of both sectors for efficiency and innovation [41].

Key scientific and technical barriers that must be overcome to integrate genomics into clinical care include the inherent complexities of genome science, inadequate training for healthcare providers regarding genomically informed practices, lack of standards in order to apply complex clinically based genomic information once it is available (i.e., policies need development) culturally bound policy values. Healthcare providers should know the basics of genomics and understand ethical and legal considerations when using genetic information in patient care. The latter might be realized under the guise of focused training programs and by integrating genomics into medical education curricula. Additionally, we need to develop user-friendly tools to help clinicians interpret genetics data, so patients are not required to have a more complex skill set and continue receiving the best care available [42].

Finally, solving the data standardization problem requires establishing best practices for collecting, analyzing, and reporting genomic data. Projects such as the Global Alliance for Genomics and Health (GA4GH) are trying to generate open standards that will enable data from genomic studies to be de-identified safely so patient confidentiality is protected and held across borders. Data from numerous populations can likewise be integrated into standardized databases, facilitating a better understanding of genetic variation and its role in disease and contributing to enhanced drug discoveries and personalized treatments [43].

9.13 TRENDS OF GENOMIC MEDICINE AND INNOVATIONS IN THE FUTURE

Genomic medicine is an evolving field, and over the next few years, several trends are setting up some exciting new developments in drug discovery and clinical care. The brand-new technologies on the horizon are set to further improve the precision, efficiency, and accessibility of genomic-based therapies, which would open up even more opportunities for treating complex and rare diseases.

9.13.1 MULTI-OMICS INTEGRATION

One of the key trends in genomic medicine is likely to involve multi-omics approaches (that integrate genomics, transcriptomes, proteomic, and metabolite data). The genomics layer captures DNA sequences; other omics provide information on gene expression, protein interactions, and metabolic pathways. Fusing these variant datasets enables researchers to better understand the molecular causes that underlie diseases and more accurately predict biomarkers or targets. One example would be merging genetic data with proteomic and metabolomic profiles to forecast patients' drug responses, promoting personalized medicine [44].

9.13.2 GENE EDITING AND SYNTHETIC BIOLOGY IMPROVEMENTS

CRISPR technology has revolutionized gene editing, and continued developments will likely broaden its applicability. In recent years, more advanced gene-editing tools that make precise changes to DNA without off-target effects have been developed or are in development for use; these include base editors and prime editing. This could mean that genetic mutations responsible for inherited diseases may finally be correctible, allowing us to cure conditions that were thought untreatable. As synthetic biology matures, developing novel biological systems, including easing synthetic genes and genomes, is just as important as gene editing. This could eventually lead to new treatments like engineered cells that recognize and kill cancer or deliver therapeutic molecules directly to specific tissues [45].

9.13.3 ARTIFICIAL INTELLIGENCE (AI) AND DATA LEARNING

Genomic medicine is on the rise, with embryo selection at IVF clinics just one example of how AI and ML are used. A practical example is using AI/ML algorithms to analyze vast, interconnected data sets, in this case, the human genome, which you can cluster together as parts of patterns and, therefore, predict how a drug would work on all individuals who contained that mutation. This might mean more sophisticated predictive models for drug interactions and patient outcomes in the future, speeding up drug discovery without losing accuracy. AI can also stratify patients based on their genetic profiles via an outpatient setting to offer personalized treatments [46]. However, as AI and ML models continue to advance in their capabilities, they are imagined playing a crucial part in customized medicine by revealing new insights that were not within the scope of traditional data analysis techniques.

9.14 GENOMIC DATABASE EXPANSION AND GLOBAL COLLABORATION

In the future, a more comprehensive range of complex and diverse ethnic populations, hence allowing for broader applications, should be utilized to develop an increasingly comprehensive genomic database. These databases are growing with the possibility of surer genetic diversity insights, which will help researchers design global medication strategies. International collaborations will also be key for disseminating data and access to resources and knowledge that facilitate the identification of new drug targets and treatments. This underscores the need for moral data sharing and normalization between regions of dissimilar foundations [47].

9.15 RNA-BASED TREATMENTS AND GENE THERAPIES

The increasing number of delivery systems and gene-editing technologies are paving the way for many more gene therapies and RNA-based treatments. This approach to RNA therapeutics has recently received widespread attention due to the success of mRNA vaccines in combating COVID-19. It is being investigated for various diseases beyond infectious disease vaccinations. Not far behind are gene therapies that edit, replace, or silence defective genes. Several such treatments have already been approved to treat disorders like spinal muscular atrophy and inherited retinal diseases [48]. A refined era of delivering these therapies safely and effectively will likely be seen in the future, leading to a decrease in immune responses and reducing off-target effects.

9.16 CONCLUSION

Incorporating genomics within drug discovery and clinical medicine emerges as a vital transformation in how diseases are conceived, diagnosed, and treated. In this chapter, we have touched upon genomics, which contributes hugely to disease pathway identification and drug discovery, personalized medicine, and the technological advancements that aid these progresses. Using genomic data, researchers, and healthcare providers can better combat these complex diseases by targeting the root causes that will translate into more personalized treatment plans. Genomics travels a long way from the lab to clinical practice with significant innovations. The chapter started with an introduction to the impact of genomics on drug discovery, emphasizing disease mechanisms and new drug targets. DNA sequencing, GWAS, and CRISPR gene editing are just some of the tools that have pushed drug development forward by allowing scientists to get a glimpse into the genetic underpinnings of diseases and precisely intervene. Genomic databases, high-throughput screening, and advances in bioinformatics all support the easy analysis of potential drug compounds.

Personalized medicine, where treatments are personalized to an individual's genetic signature, is another exciting development due to genomics. In personalized medicine, such as pharmacogenomics, drug responses can be predicted, and possible side effects can be found for optimum therapy. This technique seems to work well in many contexts, such as cancer and genetic diseases, for which treatments are of limited effectiveness. Examples, such as the use of HER2-targeted therapies in breast cancer or treatments for cystic fibrosis, illustrate this point through real-world implications based on molecular association.

In the current scenario, technologies such as NGS, CRISPR, and artificial intelligence have played a key role in popularizing genomic medicine. NGS has decreased the time and cost of sequencing and improved accuracy for genomics at scale. The advent of CRISPR and other gene editing tools creates new hope for treating disorders caused by a single known genetic change, which cannot be treated any other way. Regarding drug discovery, adding artificial intelligence and machine learning has only taken their predictive modeling abilities further than ever, enabling responses or resistance patterns. While there are significant advantages, introducing genomics in drug discovery and clinical pathways has problems. Substantial roadblocks exist to mitigating these concerns, including ethical and regulatory issues such as data privacy and consent. Moreover, technical constraints and excessive costs prevent the spread of personalized medicine. However, work is ongoing to overcome the challenges these issues represent, with research and international collaborations forging a path towards an agile healthcare ecosystem.

9.17 FUTURE DIRECTIONS

As genomic medicine advances, there are multiple areas in which more research and innovation will be necessary. Breaking down a legal and ethical wall: One of the most urgent problems is how genetic data can be safely and ethically employed. Creating solid and clear regulations will protect patient privacy and facilitate research. This will necessitate a partnership between governments, healthcare providers, and research organizations to establish enabling standards that enhance ethical behaviors and do not suppress innovation. Furthermore, we also plot the landscape for integrating multi-omics data in future studies. While genomics can provide insight, integration with transcriptomics, proteomics, and metabolomics will allow for a more complete understanding of biological processes, leading to the identification of novel biomarkers and therapeutic targets. New research needs to be conducted that improves the data integration of these complex datasets in a meaningful way to enable actionable insights. The future of genomics will be propelled by technological advancements in our ability to access genomes. Advances in gene editing, such as base editing and prime editing, may enable more precise and safer use of these new treatments. As artificial intelligence and machine learning progress, the power of genomic data analysis will be coupled to improve its predictive abilities, unearthing new drug candidates and delivering more precise predictions of how individual patients are likely to respond to treatment.

The way forward in drug discovery and clinical medicine is more genomics. Technological progress shows no sign of reaching its limits. The possibilities for what genomics can achieve in healthcare are infinite, assuming that it is founded upon proper ethical principles. It works to make medicines more personalized by using data about patients' genes and genomes rather than relying on "one size fits all" wrong assumptions. This future requires international cooperation, more work and effort in the research domain, and an inclination to meet pending difficulties head-on by all involved, which, in the end, could alter the course of medicine.

KEYWORDS

- **clinical medicine**
- **drug discovery**
- **gene editing**
- **genomic innovation**
- **high-throughput screening**
- **next-generation sequencing**
- **pharmacogenomics**

REFERENCES

1. Singh, B., Singh, A., Kaunert, C., Arora, M. K., Lal, S., & Ravesangar, K. (2025). Unleashing the ethical and legal implications of e-business revolution: Consumer privacy and security concerns in phase of digital disruption. In H. Taherdoost, G. Drazenovic, M. Madanchian, I. Khan, & O. Arshi (Eds.), *Business Transformation in The Era of Digital Disruption* (pp. 157–180). IGI Global. https://doi.org/10.4018/979-8-3693-7056-8.ch006 (accessed on 5th January 2026).
2. Singh, B., & Kaunert, C. (2025). Innovative business models in digital and virtual reality for the gig economy: Space-age sustainable development projecting smart business practices. In H. Taherdoost, G. Drazenovic, M. Madanchian, I. Khan, & O. Arshi (Eds.), *Business Transformation in The Era of Digital Disruption* (pp. 127–156). IGI Global. https://doi.org/10.4018/979-8-3693-7056-8.ch005 (accessed on 5th January 2026).
3. Singh, B., Malviya, R., & Kaunert, C. (2024). Elevating workplace sustainability for employees lensing mental health advancements: Runway for future ready healthcare services projecting SDG 3 (good health and well-being). In E. Shaikh (Ed.), *Impact of Corporate Social Responsibility on Employee Wellbeing* (pp. 285–310). IGI Global. https://doi.org/10.4018/979-8-3693-3470-6.ch014 (accessed on 5th January 2026).
4. Singh, B., Kaunert, C., & Malviya, R. (2024). Featuring machine learning in municipal solid waste management: Technological advancements in waste sorting and processing ropeway for SDG 13 (climate action). In R. Mehra, P. Kumar, & B. Hung (Eds.), *Municipal Solid Waste Management and Recycling Technologies* (pp. 343–366). IGI Global. https://doi.org/10.4018/979-8-3693-4054-7.ch013 (accessed on 5th January 2026).
5. Singh, B., Kaunert, C., & Jermsittiparsert, K. (2024). Environment-biodiversity protection and SDG 15 (life on land): Legal landscapes and probable resolutions. In B. Singh, C. Kaunert, K. Vig, & S. Dutta (Eds.), *Maintaining a Sustainable World in the Nexus of Environmental Science and AI* (pp. 259–282). IGI Global. https://doi.org/10.4018/979-8-3693-6336-2.ch010 (accessed on 5th January 2026).

6. Singh, B., & Kaunert, C. (2024). Transforming smart city and smart sustainable tourism projecting artificial intelligence and IoT: Immersive potential for revolutionizing urban living and enhancing travel experience. In K. Jermsittiparsert & P. Suanpang (Eds.), *Technological and Managerial Approaches to Fostering Sustainable Travel* (pp. 55–80). IGI Global. https://doi.org/10.4018/979-8-3693-8115-1.ch003 (accessed on 5th January 2026).
7. Raghav, A., Singh, B., Jermsittiparsert, K., Raghav, R., & Yadav, U. (2024). Artificial intelligence in environmental and climate changes: A sustainable future. In B. Singh, C. Kaunert, K. Vig, & S. Dutta (Eds.), *Maintaining a Sustainable World in The Nexus of Environmental Science and AI* (pp. 485–506). IGI Global. https://doi.org/10.4018/979-8-3693-6336-2.ch019 (accessed on 5th January 2026).
8. Kaushik, T. K., Ravish, S., Singh, A., Raghav, A., Singh, B., Arora, M. K., & Lal, S. (2024). Untapping artificial intelligence in environmental and sustainable progression: Projecting legal and ethical considerations. In B. Singh, C. Kaunert, K. Vig, & S. Dutta (Eds.), *Maintaining a Sustainable World in The Nexus of Environmental Science and AI* (pp. 109–130). IGI Global. https://doi.org/10.4018/979-8-3693-6336-2.ch005 (accessed on 5th January 2026).
9. Singh, B., Kaunert, C., Malviya, R., Lal, S., & Arora, M. K. (2024). Scrutinizing consumer sentiment on social media and data-driven decisions for business insights: Fusion of artificial intelligence (AI) and business intelligence (BI) foster sustainable growth. In A. Natarajan, M. Galety, C. Iwendi, D. Das, & A. Shankar (Eds.), *Intersection of AI and Business Intelligence in Data-Driven Decision-Making* (pp. 183–210). IGI Global. https://doi.org/10.4018/979-8-3693-5288-5.ch007 (accessed on 5th January 2026).
10. Hentze, H., Mazzitelli, I., Grosse, J., et al. (2010). Mesenchymal stem cells promote the regeneration of injured skeletal muscle by enhancing the proliferation and differentiation of satellite cells during muscle repair. *Stem Cells and Development, 19*(5), 635–645.
11. Chen, Y. S., Chen, Y. H., Cheng, Y. H., et al. (2013). Bone marrow-derived mesenchymal stem cells enhance wound healing through secretion of growth factors and cytokines. *Journal of Tissue Engineering and Regenerative Medicine, 7*(12), 883–890.
12. Lee, J. H., Lee, S. H., Lee, Y. S., et al. (2014). The potential therapeutic effects of adipose-derived stem cells in treating osteoarthritis. *Stem Cells Translational Medicine, 3*(12), 1430–1441.
13. Karp, J. M., & Leng Teo, G. S. (2009). Mesenchymal stem cell homing: The devil is in the details! *Cell Stem Cell, 4*(3), 206–216.
14. Nussbaum, S. R., Dyer, M. A., & Cagan, R. L. (2015). The role of cancer stem cells in tumorigenesis, metastasis, and therapy resistance. *Cancer Research and Treatment, 47*(2), 181–191.
15. Wang, Y., Zhang, H., Zhang, X., et al. (2022). Stem cell-based therapy for human diseases. *Signal Transduction and Targeted Therapy, 7*(1), 1–20.
16. Liu, Y., Zhang, W., Yang, Y., et al. (2020). The role of exosomes derived from mesenchymal stem cells in liver disease. *Frontiers in Pharmacology, 11*, 578.
17. Zhao, T., Zhang, Z. N., Westenskow, P. D., et al. (2015). Human iPS cell-derived RPE cells induce retinal regeneration and restore vision in retinal degeneration. *Cell Stem Cell, 16*(1), 70–83.

18. Saha, S., Jaeger, M., Hwang, H., et al. (2020). The role of miRNA in regulating stem cell differentiation and function. *Frontiers in Genetics, 11*, 123.
19. Ghosh, D., Ghosh, S., Nandi, A., et al. (2020). The role of mesenchymal stem cells in tissue engineering and regenerative medicine. *Current Stem Cell Research & Therapy, 15*(4), 294–306.
20. Li, Y., Xu, C., Jiang, H., et al. (2020). The potential application of induced pluripotent stem cells in regenerative medicine. *Frontiers in Cell and Developmental Biology, 8*, 167.
21. Choi, S. H., Kim, H. J., Kim, Y. S., et al. (2020). The role of exosomes derived from mesenchymal stem cells in promoting wound healing through their anti-inflammatory properties enhancing angiogenesis, fibroblast proliferation, collagen deposition, reepithelialization, granulation tissue formation, etc. *Frontiers in Pharmacology, 11*, 578. https://doi.org/10.3389/fphar.2020.00578 (accessed on 5th January 2026).
22. Tzeng, Y. C., Chen, T. H., Huang, C. C., et al. (2020). The potential application of mesenchymal stem cells for treating various diseases including cardiovascular diseases, neurodegenerative diseases, autoimmune disorders, etc. *Stem Cells Translational Medicine, 9*(3), 322–333. https://doi.org/10.1002/sctm.19-0437 (accessed on 5th January 2026).
23. Huang, X., Xu, D., Wang, Z., et al. (2020). The role of exosomes derived from mesenchymal stem cells in promoting wound healing through their anti-inflammatory properties enhancing angiogenesis, fibroblast proliferation, collagen deposition, reepithelialization, granulation tissue formation, etc. *Frontiers in Pharmacology, 11*, 578. https://doi.org/10.3389/fphar.2020.00578 (accessed on 5th January 2026).
24. Wu, Y., Chen, L., Scott, P. G., et al. (2020). Mesenchymal stem cells promote wound healing through paracrine mechanisms involving growth factors, cytokines, extracellular vesicles, etc. *Journal of Tissue Engineering and Regenerative Medicine, 14*(3), 424–434. https://doi.org/10.1002/term.3026 (accessed on 5th January 2026).
25. Lee, J. H., Lee, S. H., Lee, Y. S., et al. (2014). The potential therapeutic effects of adipose-derived stem cells in treating osteoarthritis through their anti-inflammatory properties promoting cartilage regeneration, reducing pain, improving joint function, etc. *Stem Cells Translational Medicine, 3*(12), 1430–1441. https://doi.org/10.5966/sctm.2014-0089 (accessed on 5th January 2026).
26. Zhao, T., Zhang, Z. N., Westenskow, P. D., et al. (2015). Human iPS cell-derived RPE cells induce retinal pigment epithelium regeneration *in vivo* and restore vision in a rat model of retinal degeneration. *Cell Stem Cell, 16*(1), 70–83. https://doi.org/10.1016/j.stem.2014.11.018 (accessed on 5th January 2026).
27. Saha, S., Jaeger, M., Hwang, H., et al. (2020). The role of miRNA in regulating stem cell differentiation and function during development and disease processes including cancer, neurodegenerative diseases, metabolic disorders, etc. *Frontiers in Genetics, 11*, 123. https://doi.org/10.3389/fgene.2020.00123 (accessed on 5th January 2026).
28. Ghosh, D., Ghosh, S., Nandi, A., et al. (2020). The role of mesenchymal stem cells in tissue engineering, regenerative medicine applications including bone, cartilage, skin, nerve, heart, liver, pancreas, etc. *Current Stem Cell Research & Therapy, 15*(4), 294–306. https://doi.org/10.2174/1574888X15666200110113 5723 (accessed on 5th January 2026).

29. Li, Y., Xu, C., Jiang, H., et al. (2020). The potential application of induced pluripotent stem cells in regenerative medicine for various diseases including heart disease, diabetes, neurodegenerative diseases, etc. *Frontiers in Cell and Developmental Biology, 8*, 167. https://doi.org/10.3389/fcell.2020.00167 (accessed on 5th January 2026).
30. Gavasso, S., Kråkenes, T., Olsen, H., Evjenth, E. C., Ytterdal, M., Haugsøen, J. B., & Kvistad, C. E. (2024). The therapeutic mechanisms of mesenchymal stem cells in MS—A review focusing on neuroprotective properties. *International Journal of Molecular Sciences, 25*(3), 1365. https://doi.org/10.3390/ijms25031365 (accessed on 5th January 2026).
31. Ma, C. Y., Zhai, Y., Li, C. T., Liu, J., Xu, X., Chen, H., Tse, H. F., & Lian, Q. (2024). Translating mesenchymal stem cell and their exosome research into GMP-compliant advanced therapy products: Promises, problems and prospects. *Medicinal Research Reviews, 44*(3), 919–938. https://doi.org/10.1002/med.22094 (accessed on 5th January 2026).
32. Shi, L., Chen, L., Gao, X., Sun, X., Jin, G., Yang, Y., Shao, Y., Zhu, F., & Zhou, G. (2024). Comparison of different sources of mesenchymal stem cells: Focus on inflammatory bowel disease. *Inflammopharmacology*, 1–22. https://doi.org/10.1007/s10787-024-01373-1 (accessed on 5th January 2026).
33. Surico, P. L., Scarabosio, A., Miotti, G., Grando, M., Salati, C., Parodi, P. C., Spadea, L., & Zeppieri, M. (2024). Unlocking the versatile potential: Adipose-derived mesenchymal stem cells in ocular surface reconstruction and oculoplastics. *World Journal of Stem Cells, 16*(2), 89. https://doi.org/10.4252/wjsc.v16.i2.89 (accessed on 5th January 2026).
34. Cerneckis, J., Cai, H., & Shi, Y. (2024). Induced pluripotent stem cells (iPSCs): Molecular mechanisms of induction and applications. *Signal Transduction and Targeted Therapy, 9*(1), 112. https://doi.org/10.1038/s41392-024-01789-4 (accessed on 5th January 2026).
35. Ghoneim, M. A., Gabr, M. M., El-Halawani, S. M., & Refaie, A. F. (2024). Current status of stem cell therapy for type 1 diabetes: A critique and a prospective consideration. *Stem Cell Research & Therapy, 15*(1), 23. https://doi.org/10.1186/s13287-024-03691-8 (accessed on 5th January 2026).
36. Farokhi, S., Razavi, Z. S., Mavaei, M., Shadab, A., Afkhami, H., & Sardarabadi, H. (2024). New perspectives on arteriosclerosis treatment using nanoparticles and mesenchymal stem cells. *Discover Applied Sciences, 6*(8), 1–40. https://doi.org/10.1007/s42452-024-06488-7 (accessed on 5th January 2026).
37. El Masri, J., Fadlallah, H., Al Sabsabi, R., Afyouni, A., Al-Sayegh, M., & Abou-Kheir, W. (2024). Adipose-derived stem cell therapy in spinal cord injury. *Cells, 13*(17), 1505. https://doi.org/10.3390/cells13171505 (accessed on 5th January 2026).
38. Mamachan, M., Sharun, K., Banu, S. A., Muthu, S., Pawde, A. M., Abualigah, L., & Maiti, S. K. (2024). Mesenchymal stem cells for cartilage regeneration: Insights into molecular mechanism and therapeutic strategies. *Tissue and Cell, 82*, 102380. https://doi.org/10.1016/j.tice.2024.102380 (accessed on 5th January 2026).
39. Mahdi, M. S. (2024). Clinical applications of stem cell therapy in the fight against cancer. *Journal of Medical Genetics and Clinical Biology, 1*(8), 188–199.

40. Baig, M. S., Ahmad, A., Pathan, R. R., & Mishra, R. K. (2024). Precision nanomedicine with bio-inspired nano systems: Recent trends and challenges in mesenchymal stem cells membrane-coated bioengineered nanocarriers in targeted nanotherapeutics. *Journal of Xenobiotics, 14*(3), 827–872. https://doi.org/10.3390/jox14030061 (accessed on 5th January 2026).
41. Garcia, S. G., Clos-Sansalvador, M., Sanroque-Muñoz, M., Pan, L., & Franquesa, M. (2024). Functional and potency assays for mesenchymal stromal cell-extracellular vesicles in kidney disease. *Current Opinion in Physiology, 39*, 100746. https://doi.org/10.1016/j.cophys.2024.100746 (accessed on 5th January 2026).
42. Niu, Y., Ji, J., Yao, K., & Fu, Q. (2024). Regenerative treatment of ophthalmic diseases with stem cells: Principles, progress, and challenges. *Advances in Ophthalmology Practice and Research, 4*, 100091. https://doi.org/10.1016/j.aopr.2024.100091 (accessed on 5th January 2026).
43. Abdulmalek, O. A., Husain, K. H., AlKhalifa, H. K., Alturani, M. M., Butler, A. E., & Moin, A. S. (2024). Therapeutic applications of stem cell-derived exosomes. *International Journal of Molecular Sciences, 25*(6), 3562. https://doi.org/10.3390/ijms25063562 (accessed on 5th January 2026).
44. Song, J., Zhou, D., Cui, L., Wu, C., Jia, L., Wang, M., Li, J., Ya, J., Ji, X., & Meng, R. (2024). Advancing stroke therapy: Innovative approaches with stem cell-derived extracellular vesicles. *Cell Communication and Signaling, 22*(1), 369. https://doi.org/10.1186/s12964-024-01499-z (accessed on 5th January 2026).
45. Piao, X., Li, D., Liu, H., Guo, Q., & Yu, Y. (2024). Advances in gene and cellular therapeutic approaches for Huntington's disease. *Protein & Cell*. https://doi.org/10.1093/procel/pwae042 (accessed on 5th January 2026).
46. Yuan, H. L., Chang, L., Fan, W. W., Liu, X., Li, Q., Tian, C., Zhao, J., Li, Z. A., Pan, X. H., & Zhu, X. Q. (2024). Application and challenges of stem cells in cardiovascular aging. *Regenerative Therapy, 25*, 1–9. https://doi.org/10.1016/j.reth.2024.03.001 (accessed on 5th January 2026).
47. Lin, H., Zhou, J., Ding, T., Zhu, Y., Wang, L., Zhong, T., & Wang, X. (2024). Therapeutic potential of extracellular vesicles from diverse sources in cancer treatment. *European Journal of Medical Research, 29*(1), 350. https://doi.org/10.1186/s40001-024-01421-7 (accessed on 5th January 2026).
48. Johnson, V. A. (2024). Review of current and potential applications of mesenchymal stem cells in exotic animal species. *Journal of the American Veterinary Medical Association, 262*(S1), S131.

CHAPTER 10

Impact of Genomics on Drug Discovery and Clinical Medicine: Fostering Gene Editing, Personalized Medicine, Bioinformatics, and Artificial Intelligence

RISHABHA MALVIYA[1] and BHUPINDER SINGH[2,*]

[1]*Galgotias University, Greater Noida, Uttar Pradesh, India*

[2]*Sharda School of Law, Sharda University, Greater Noida, India; University of South Wales, UK; Santo Tomas University Tunja, Colombia*

*Corresponding author

ABSTRACT

The combined power of genomics in drug discovery and clinical practice is rewriting the history of healthcare with gene editing and personalized medicine. The capability to do gene editing with great precision through advancements in technology such as CRISPR is advancing apace, providing hope that it may soon see cures for genetic diseases and individualized therapies. Genomics-based personalized medicine allows treatments to be tailored according to each person's unique genetic information, for higher efficacy and lower side effects. Bioinformatics has become essential to the management and analysis of the enormous datasets produced by genomic research, which builds on bioinformatics and drives discovery in drug targets, thereby accelerating drug development. Artificial intelligence improves upon this by analyzing genomic data in real time for patterns to predict drug responses and reduce the time taken for a clinical trial.

Genetic Editing for Healthcare: Advances in Disease Management.
Bhupinder Singh, Rishabha Malviya, Sahil Lal, and Sathvik Belagodu Sridhar (Eds.)

DOI: 10.1201/9781779643827-10

Combined, these advances herald an era where therapies are no longer one-size-fits-all but individually matched, transforming both preventative and therapeutic practice in the modern age of medicine.

10.1 INTRODUCTION

The recent decades have seen extraordinary advances in genomics, reshaping the state of the art and practice of drug discovery and clinical medicine. Now, with the completion of the Human Genome Project and subsequent maturation of next-generation sequencing (NGS) technologies, decoding the DNA blueprint for human beings has gone from an entirely consuming and demanding exercise to a set of streamlined steps available to researchers and clinicians around the world [1]. A new era of doing medicine has emerged through tremendous gain in genomics knowledge that has already precipitated the shift from a formula- or heuristics-oriented approach towards much more individualized and modern therapeutic interventions driven by gene editing, bespoke feed, bioinformatics, and artificial intelligence (AI), all optimized for our practice domain [2].

The notion of personalized medicine, applying genomic data to tailor treatments to individual patients based on their unique genetic profile, is a radical departure from the "one-size-fits-all" approach of traditional medicine, a model that could potentially better efficacy at lower adverse effects [3]. Meanwhile, gene-editing technologies such as CRISPR-Cas9 have opened new avenues for modifying DNA at an unparalleled level of precision, which opens new pathways for treating genetic disorders and offers the possibility to completely eradicate Mendelian diseases [4]. Through the bioinformatics and artificial intelligence integration in genomics, more advancements are rapidly coming in. Bioinformatics helps us to manage and interpret the huge data generated by genome studies, while AI is used for faster and more accurate data analysis, drug target identification, and predictive modeling [5]. The combination of these technologies provides deeper insights into the genetic determinants of disease, which enables more informed drug discovery and clinical decision-making.

It takes a deeper look at these areas and the influence of genomics on drug research and clinical medicine in general, with particular emphasis on gene editing, personalized medicine, bioinformatics, and AI [6]. Through exploring these interrelated domains, our objective is to underscore the transformative power of genomics in healthcare as a whole, with improved patient results and advances in innovative targeted therapeutic modalities (Figure 10.1) [7].

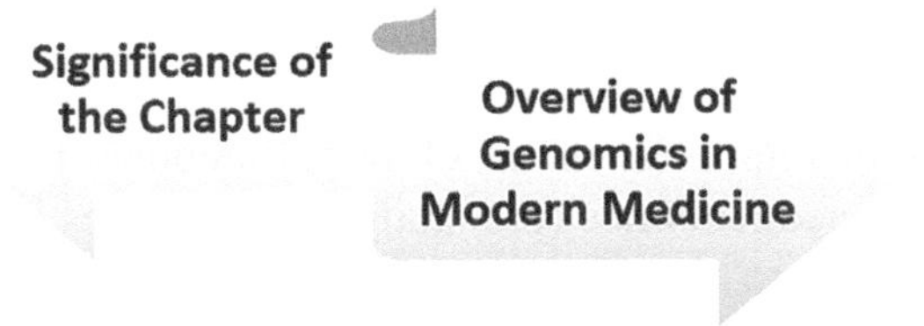

FIGURE 10.1 Dimensions of introduction section.

Source: Self-Prepared – Original.

10.1.1 SIGNIFICANCE OF THE CHAPTER

The increasing body of genomic disease research has broad ramifications for drug discovery and clinical medicine, rendering this chapter particularly germane to current scientific and medical discourse. The goal is to reemphasize that genomics is a transformational force, rewriting our understanding of disease, diagnostics, and therapeutics. This is where the potential of genomics, combined with gene-editing, bioinformatics, and AI technologies coming together, to empower healthcare and offer resolutions which were earlier deemed as unachievable such as [8]:

1. **Drug Discovery:** Transforming a traditionally slow, expensive, and risky drug discovery process. With traditional methods, understanding genetics is a trial-and-error process that can be slow because there are inherent genetic differences between patients. The new genomics-based era in drug discovery is a good example of how such genetic information can be exploited for the identification of specific molecules and pathways that play a role in disease processes [9]. Understanding the biology, or genetic basis, of disease can enable researchers to create drugs that have precise biological mechanisms so that clinical trial success rates rise and development time is reduced. Also, by identifying biomarkers that can predict how patients will respond to a given therapy based on their genomics, medicine developers are able to produce more efficacious drugs with a greater safety profile [10].

2. **Promoting Personalized Medicine:** Personalized medicine is one of the greatest emerging implications of genomics. Clinicians can use genomic data to individualize medical treatment based on an individual's genetic profile, lifestyle, and environmental factors. This more personalized form of medicine could provide improved therapeutic efficacy, a reduction in adverse drug events, and an improvement in clinical outcomes for the patients [11]. Pharmacogenomics (the study of how genes influence a person's response to drugs) has already resulted in the development of existing gene-based tests that help determine if you take far too little or too much medication (as well as which one will work or stop working) for therapies in cancers, cardiovascular diseases, and psychiatric conditions. Personalized medicine: a shift from reactive treatment designed to react to root causes of disease toward preventive healthcare, finely tuned for the genetic susceptibilities of each individual.
3. **Gene Editing Pipeline:** Gene editing is an area that has captured the attention of both scientific and popular press, CRISPR-Cas9 in particular, for its potential to cure genetic diseases at their source [12]. These molecular scissors are a highly advanced technology that allows scientists to cut and paste the DNA of any living organism, thereby giving hope for curing inherited diseases such as cystic fibrosis, muscular dystrophy, or sickle cell anemia. Since gene editing fixes the genetic mutation itself, it offers a potential long-term cure over and beyond symptom alleviation. Further, gene editing applications could fuel the discovery of new therapies to address a previously intractable disease, thereby broadening the potential for clinical medicine.
4. **Data Analysis Boost with Bioinformatics and Artificial Intelligence (AI):** As data generated from all the Genomic research is huge, they need a tool to analyze, which becomes bioinformatics and thereafter AI. Bioinformatics, as a computational discipline, provides the tools to structure and analyze genomic data in order to unveil patterns and relationships that would go unnoticed otherwise. Additionally, AI improves the entire process by allowing the rapid and precise examination of data sets using machine learning algorithms for drug response predictions as well as potential

therapeutic target identification. AI-driven models can emulate intricate biological systems, conduct clinical trials development, and identify a suitable candidate to be put on trial, hence making the entire drug discovery process cost-effective. In healthcare, AI helps to create diagnostic tools that predict if a disease will develop from a patient's genomics, thus allowing for an early intervention and personalized treatment plans which are geared toward the individual [13].

The role genomics will play in transforming drug discovery and clinical practice is essential to the promise of a new era in healthcare innovation. The merging of gene editing, personalized medicine, bioinformatics, and AI is rapidly transforming the treatment landscape, offering the tools required to address complex medical dilemmas at their root. Future medicine, benefiting from the unions of genomic understanding and cutting-edge technologies, one day will deliver truly state-of-the-art efficacious patient-centric targeted therapies, saving and enhancing lives across the globe [14].

10.1.2 OVERVIEW OF GENOMICS IN MODERN MEDICINE

By illuminating the factors that influence health and disease, genomics has fundamentally transformed modern medicine. New technologies, such as next-generation sequencing, mean researchers are able to map complete genomes faster and cheaper than ever before, leading to advances in how we diagnose, treat, and prevent disease [15]. Genomics is one of the major forces behind moving from traditional, strictly generalized healthcare to more personalized, precise health care in modern medicine. By identifying the genetic differences that underpin disease susceptibility and response to treatment, genomics will enable targeted therapies, particularly in oncology, cardiovascular disease, and rare genetic disorders. With genomics, personalized medicine treats the patient with their unique genetic background, which increases survival times and decreases adverse effects in treatments.

Genomics has also opened our knowledge of complex diseases by spotting single nucleotide polymorphisms or biomarkers, which could test how far a disease has progressed and what drug would be effective. For example, gene editing technologies (e.g., CRISPR-Cas9) offer the possibility to correct genetic deficits and thus cure hereditary diseases. As part and parcel of an interdisciplinary approach that also includes bioinformatics and deep learning, genomics is enabling better diagnostics, allowing rapid drug discovery processes, and the creation of novel gene-based therapeutics, heralding a new age of medical science [16].

10.2 GENOMICS AND ITS ROLE IN DRUG DISCOVERY: GENOMICS-DRIVEN DRUG TARGET IDENTIFICATION

Genomics has emerged as an important tool in the pharmaceutical sector with the potential to dramatically quicken discovery of novel drug targets, facilitate understanding of pathways involved in disease formation, and help develop more sophisticated therapeutics. Genomics is the field of science that involves mapping and understanding thc structure, function, and evolution of genomes, illuminating our genetic blueprint and shedding light on disease genetics. This type of information is critical for discovering drug targets: molecules in the body (typically proteins) whose functions could be modulated by a drug to impact disease. Identification of drug targets through genomics has catalyzed a new, faster, more holistic era in drug development – achieving what previously were low-yielding and serendipitous drugs.

Historically, drug discovery was hinged on empiricism such as trial-and-error, chemical screening, or hypotheses based on limited knowledge of disease mechanisms. The advent of the genomic era has been made possible by the completion of the Human Genome Project in 2003 and rapid advances in high-throughput sequencing technologies, which have resulted in an explosion of genetic data. For research, whole genomes or the entirety of exomes, the portions of an organism's genome that encode protein-coding gene products, can be analyzed to learn about mutations, variations, and alterations related to particular diseases. With the help of such genomic data, scientists come to know about how diseases work at a molecular level and what are the targets for therapeutic intervention. As an example, if researchers sequence the genomes of cancer patients, they can figure out what sort of mutations are causing the tumor to grow, and those specific mutated genes or proteins become good targets for drugs. This has given us breakthroughs for cancer-related targets, such as BRAF mutation in prostate or EGFR mutation in lung, and has made the rise of new therapies possible: targeted therapy [17].

Genomics-driven drug target discovery involves harnessing information derived from genomics to discover novel molecular targets underlying disease. This process begins with the identification of either genes or genotypes that are related to disease susceptibility, progression, or drug response [18]. Once identified, these genes can be further studied to understand their biological role and if they are a druggable target. The main methods are genome-wide association studies (GWAS), where you compare whole genomes from an entire population of affected individuals to that of non-affected ones. This way, researchers can determine genetic differences between the two groups and identify genetic variations that increase the risk of disease. Many of these variants fall in or near genes encoding proteins involved in disease processes, thereby opening up novel hypotheses for drug targets.

Once a genetic variant or mutation has been linked to a disease, functional genomics techniques such as RNA sequencing and CRISPR gene editing will help us study the way in which those genes and proteins contribute to various biological pathways. It helps determine if modification of the actions for those genes or proteins is in turn beneficial for the treatment of that disease. For example, CRISPR-Cas9 has helped to validate drug targets by providing a method for turning on or off genes in cellular models of disease and experimental animals. The other application of genomics is in drug target identification, where it can reveal potential avenues for treatment that would have been previously impossible to discover. In like manner, genomics also helps with biomarker discovery, which are the life-blood of many personalized medicine programs. Biomarkers are biological markers that can indicate what might happen to a disease, such as different proteins or gene expressions. Biomarkers are used to stratify patients into subgroups based on their genetic profiles, resulting in personalized treatments. In medicine, genomics-guided biomarker discovery fuels targeted therapies such as trastuzumab (Herceptin) that are exclusively for breast cancer patients with HER2-positive tumors [19].

The first advantage is that genomics provides a more logical and evidence-based way of identifying drug targets, as opposed to the old trial-and-error approach. Starting from already identified causes of genetic mutations or alterations is more likely to produce a functionally relevant target. This enhances the success of drug development by targeting components that are necessary for advancing the disease [20]. In a second way, genomics can help identify drug targets for diseases with complex genetic underpinnings, For example, Alzheimer's disease, cardiovascular disorders, and autoimmune diseases. Most of these diseases are complex in that they involve multiple genes and pathways, and genomics is helping to identify the key determinants underlying those processes, shining a light on specific genes or proteins that have a prominent role in disease pathogenesis. Genomics-driven drug discovery has its advantages but also poses several challenges. Most diseases are polygenic, meaning risk comes from more than one gene, and so target identification is harder. Moreover, not all genetic mutations are "druggable," For example, they do not encode proteins to which a conventional small-molecule drug can be directed. However, translating genomic discoveries to therapeutic utility is slow and costly in terms of the genomics research itself [21].

The landscape of drug discovery has been fundamentally transformed by genomics, which provides greater insight into the molecular underpinnings of disease. This new genomics-driven drug target identification has enabled more completely grounded, therapeutically compelling solutions to pinpoint particular loci where variations and mutations present the opportunity for effective therapeutic intervention, particularly in diseases such as oncology, rare genetic disorders, and personalized medicine. Although challenges persist, the combination of genomics with other advanced science and technologies, like bioinformatics and artificial intelligence, has the potential to revolutionize drug discovery for years to come (Figure 10.2) [22].

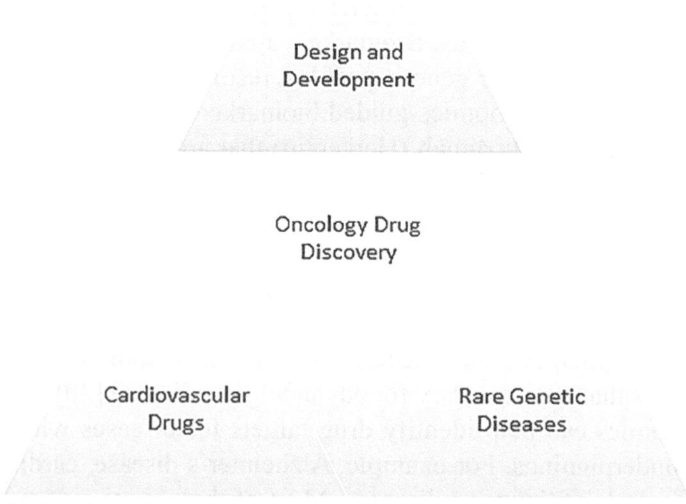

FIGURE 10.2 Genomics in successful drug discovery.

Source: Self-Prepared – Original.

10.3 GENE EDITING IN MEDICINE: REVOLUTIONIZING THERAPEUTIC STRATEGIES

Gene editing has become a game-changer for the treatment approaches in modern medicine, providing unprecedented opportunities to cure genetic diseases and complex disorders, even contributing to preventative health care. It is a cutting-edge technology that enables any DNA sequence or gene of an organism, including humans, at any position and in elaborate ways without leaving any trace (unlike other systems/products available today). The gene-editing tool that has most caught the imagination of therapeutic scientists is CRISPR-Cas9, which makes it possible to make targeted modifications in the genome with a precision and efficiency previously unthinkable.

Gene editing is the alteration of gene cells directly at their source, which means you can change the genetic make-up of genes. This may include the removing, adding, or changing of particular genetic materials to create the intended therapeutic result. Gene editing allows scientists to treat the causes of many diseases, especially those that result from a mutation in a single gene: sickle cell anemia, cystic fibrosis, muscular dystrophy, or hemophilia [23]. Current treatments for these diseases usually provide only symptomatic relief, not treatment of the root genetic cause, but gene editing provides the possibility of a one-time corrective intervention. CRISPR-Cas9 is one of the most popular gene-editing techniques, and has emerged in a variety of ways since its easy design, high efficiency, and precise ability to edit DNA sequence were realized. Other methods like TALENs (transcription activator-like effector nucleases) and ZFNs (zinc-finger nucleases) have been used too, but CRISPR has surpassed them mainly because of its speed and accuracy in locating, cutting, and replacing DNA sequences, which makes it a revolutionary tool for gene editing in medicine.

In medicine, gene editing's single biggest benefit is providing exact medications that turn off the genetic defects causing illnesses. This is especially revolutionary in genetic disorders with a known gene mutation. The most advanced realization for gene editing is illustrated by the development of hematologic diseases, particularly debilitating sickle cell and beta-thalassemia. The two conditions are both monogenic, and early clinical trials aimed at editing a patient's hematopoietic stem cells (HSCs) using CRISPR technology have been very encouraging. This ability to repair the faulty genes in these cells means gene editing could offer a very real cure for conditions previously only treatable through life-long drug regimens and dangerous procedures such as bone marrow transplants. In addition to blood diseases, gene editing also holds promise as a means to treat cancer and can be used in the development of therapies that improve the body's immune response against cancer cells. This technique, known as CAR-T cell therapy, genetically reconfigures an individual's T cells in a way that they'll target their weapons more precisely to fight the disease. It is already being used for particular kinds of blood cancers. Gene editing could also be used to improve such therapies even more, either enhancing the specificity of immune cells or mitigating their potential risk for off-target adverse events.

This technique could provide a means of treating inherited blindness, muscular dystrophy, and many other diseases known to be linked to certain genetic mutations. Researchers are investigating the disease, such as editing genes to fix mutations in the dystrophin gene which causes Duchenne muscular dystrophy, a deadly muscle-wasting condition. The restoration of dystrophin production would therefore represent an eagerly anticipated therapeutic option for patients with few options available to them [24].

Gene editing is additionally valuable for the improvement of a customized solution. By programming interactions between the systems, specific genetic sequences can be manipulated on a patient-to-patient basis to design treatments suited for an individual's unique genetics. This is particularly important for diseases such as cancer, where genetic mutations that drive the growth of tumors can vary from patient to patient with a given type of tumor [25]. The new era of personalized cancer therapy by *in situ* gene editing in tumor cells or *ex-vivo*-gene-editing (in immune-cells) are innovative and less-toxic methods for better therapeutic outcome than employing traditional treatments such as chemotherapy or radiation. Gene editing could be applied to the prevention of diseases before they develop. High-risk people for particular genetic diseases might, apparently eventually, have disease-causing mutations edited out of their genes before symptoms develop [26]. Using this preventive paradigm to address genetic liabilities that could put our children at risk for Alzheimer's, heart disease, cancer, and perhaps other conditions, would be nothing short of revolutionary in terms of contributing to "health" in the health care movement.

Despite the great potential, gene editing admits to a large and complex set of ethical and technical concerns. Germline editing, the genes of embryos or reproductive cells, could leave permanent modifications that would be inherited by later generations, leading to fears about unintended outcomes and the creation of designer babies. The possibility of producing genetic modifications, instead of just fixing mutations that cause diseases, also raises the ethical problem concerning fairness, authorization, and the social effects of changing human genes. Ethical guidelines and global agreements are necessary to regulate the use of gene-editing technology. Obviously, the gene-editing methods need to be safe and specific. Off-target effects, meaning editing parts of the genome where it was not supposed to, could cause additional mutations or adverse reactions in living organisms [27]. As the research advances, these critical challenges must be addressed to unlock the full potential of precise gene-editing tools and incorporate appropriate safety measures.

Gene editing has revolutionized the field of medicine by providing therapies that are accurate, specific, and potentially curative for genetic diseases and complex disorders. The advent of technologies such as CRISPR has raised the prospect not only to correct genetic defects, but also to personalize therapy on a patient-by-patient basis. There continue to be roadblocks along the way, but the potential is there for gene editing to upend standard therapeutic protocols, heralding a new era in precision medicine that could even one day wipe out certain genetic diseases [28].

10.4 PERSONALIZED MEDICINE: TAILORING TREATMENTS TO GENETIC PROFILES

Precision medicine is a model of healthcare that is tailored to genes, environment, and lifestyle factors for each person. In contrast with the "one-size-fits-all" strategy that historically has guided medical practice, personalized medicine aims to identify the genetic, molecular, and environmental underpinnings of individual variability in therapeutic response. The method is used to more precisely diagnose disease, predict the risk of developing a disease, and make it possible to prescribe treatments which are more effective than the ones currently in use, with fewer side effects [29].

In its simplest form, personalized medicine uses advances in genomics, bioinformatics, and medical technology to design treatment that is tailored to the genetic make-up of a patient. The sequencing of the human genome and advances in genomic technologies describes the advent of NGS have had a great impact on this change. DNA sequencing allows doctors to pinpoint genetic mutations or variations that cause diseases or affect how a patient processes and metabolizes drugs. This data can help tailor treatments more likely to work for that person. Nowhere has the promise of personalized medicine been more profound than in oncology, where tumor genotyping is mandated for virtually every cancer that presents. Cancer is an incredibly heterogeneous disease—every tumor has its own unique genetic fingerprints [30]. With personalized medicine, oncologists can analyze the genetic characteristics of a tumor and choose targeted therapies based on the specific mutations that promote growth in cancer. For instance, anti-HER2 therapies in breast cancer or EGFR-mutant lung cancer have provided patients with these mutations far more effective treatment options than chemotherapeutic approaches.

The hallmarks of personalized medicine are pharmacogenomics, which refers to how a person's response to medications can be affected by their genetics [31]. This can have an effect on the function of certain medications and, thus, their ability to work safely or at all (in other words, the efficacy of a drug in relation to its risk/benefit ratio), within specific groups. This information can help doctors tailor drug chemistry to provide the right amount of a drug without causing side effects [32]. For example, some genetic mutations in the CYP450 enzymes can alter how patients metabolize popular medications such as antidepressants, blood thinners, and cancer treatments. Some variants may necessitate higher or lower doses, or prefer different medications altogether. Pharmacogenomics has also been considered in cardiac care to guide the administration of blood-thinning medications such as clopidogrel so that non-metabolizer patients are not prescribed these drugs, which would prevent ineffective treatment or harm [33].

The promise of the personalized medicine future is in fact when it also merges with other types of data (beyond just genetic), such as proteomics (study of proteins), metabolomics (study of metabolites, lipids, etc.), and environmental factors. Making the largest datasets will grow even with augmented AI predictive capabilities and machine learning, but this is likely to be related to variations in disease risk, response to treatment accuracy of prediction rather than fundamental change [34]. Despite considerable challenges, such as cost, data privacy, and regulatory approval, the promise of improved patient outcomes through personalized medicine cannot be ignored. Personalized medicine, which uses individual genetic profiles to tailor treatments, is transforming healthcare and holds out the promise of more effective, safer, and more precise therapies for a multitude of diseases [35].

10.5 BIOINFORMATICS IN GENOMIC DATA MANAGEMENT AND ANALYSIS

Bioinformatics, the management and analysis of genomic data, is a subset of computational biology, which in itself is an emerging field that bridges biological research with cultural transformations going on worldwide [36]. As the volume of data continues to grow, and as more fractions of this data will be generated by an increasing number of feasible technical methods like NGS and whole-genome sequencing, there is a dramatic increase in the need for bioinformatics support that will enable the storage, organization, and interpretation of massively parallel genomics information. Genomics enables researchers, clinicians, and others to analyze raw genetic data in order to draw conclusions for personalized medicine, such as drug discovery or the diagnosis of disease [37].

Life bioinformatics is a vast and fast-growing field of science that requires managing the great amount of data generated by genomic research. Data management in genomics projects, such as the Human Genome Project or 1000 Genomes Projects, have delivered terabytes of data which then had to be stored, organized, and shared with other researchers around the world. Bioinformatics offers specialized databases such as GenBank, EMBL, and the European Bioinformatics Institute (EBI) for the storing, recovery, and sharing of extensive data [38]. They act as repositories for DNA sequences, gene annotations, and other relevant metadata so that scientists can readily access and analyze genetic information in humans or others. Furthermore, sublime computational resources like cloud and high-performance computing (HPC) are increasingly playing crucial roles in the handling of massive genomic datasets and in supporting large international collaboration over the web [39]. Bioinformatic tools provide data integrity, security, and scalability that cope with the enormous growth of genomic research.

Bioinformatics plays a critical role in deciphering genome data and gaining valuable biological insights, rather than just managing data. Genomic sequences, which may comprise tens of billions of nucleotides, are simply too large and complex to manually investigate. Bioinformatics algorithms and tools perform sequence alignment, gene annotation, variant calling, and functional analysis. There are many tools that help to align sequences; for example, BLAST and Bowtie can compare DNA or protein sequence data while searching for evolutionary relationships or gene functions [40]. These tools are especially important when trying to determine how genes are maintained throughout different species or how mutations can cause diseases. At the same time, bioinformatics pipelines supersede this by performing variant analysis and identifying SNPs, insertions/deletions, and other mutations for causing genetic disorders or drug response variations [41].

Bioinformatics helps in managing and analyzing genomic data, which is essential for researchers to work with big data taking thousands of conforming genes. Bioinformatics connects computational tools to biological data, creating a new understanding of the molecular mechanisms of diseases and uses large amounts of genomic data for personalized medicine, discovering new drugs, and prevention [42]. These functional analysis tools allow researchers to know the biological functions of related genes and their participation in cellular processes (gene ontology [GO] and pathway enrichment analyses). This type of information is crucial for finding targets for drug development and understanding how diseases develop to find new treatment options [43].

10.6 ARTIFICIAL INTELLIGENCE (AI) IN GENOMIC RESEARCH AND DRUG DISCOVERY

AI is changing the landscape of genomics research and drug discovery by using cutting-edge technology to analyze data more quickly, pick up on patterns better with increasingly high accuracy, and predict outcomes [44]. With more data being generated than ever before in the field of genomics, AI has emerged as a saving grace for the sheer volume of computational power required to process, analyze, and understand this information in a new wave of thinking about genetics that allows us to create targeted interventions [45].

In genomic research, AI is now an essential tool that helps researchers analyze large datasets faster. AI, in particular machine learning algorithms, is a branch of AI proven to detect order within genomic sequences that might relate to diseases or characteristics [46]. Their high-throughput potential means that these algorithms can sift through huge troves of genomic data, such as the sequencing data from NGS technologies, and, hopefully, locate mutations, genetic variations, and disease-associated markers faster and more accurately than with traditional methods. And deep learning, a subset of AI, is also effective in analyzing complex genomic data. This includes interpreting the function of previously unknown genes, understanding gene expression patterns, and predicting how mutations in particular genes might influence disease outcomes [47]. The GWAS that are used to identify genetic variants corresponding to various complex diseases such as cancer, diabetes, or Alzheimer's disease are also supported by AI. AI can speed up the discovery of new biomarkers and therapeutic targets by identifying relevant variants much more quickly than humans [48].

Artificial intelligence is poised to revolutionize drug discovery, streamlining everything from target identification to clinical trials. A key area in which AI has shown promise is target identification, where machine learning models are used to predict druggable candidate targets from genomic and proteomic data. AI can identify key molecular targets in disease pathways by cross-correlating gene-disease associations and protein interactions, which in turn helps develop drugs more focused and efficiently. AI also helps in drug development by predicting the behavior of small molecules and how they can interact with these targets when potential protein structures are identified to be used as drug targets [49].

AI algorithms that imitate interactions between molecules are used to predict the likely efficacy of a given compound, thereby saving time and resources in the initial phase of drug discovery. Pharmaceutical companies can thus use this computational approach to identify drugs that are safer and more effective [50]. AI is used in drug repurposing, to test existing drugs for new therapeutic uses. As an example, a previously approved drug might be effective for other diseases than the affected one based on patient data and data pattern with current technology [51]. Using AI, genomic research and drug discovery are changing with rapid and precise methods to decipher complex datasets. Using AI allows researchers to locate and identify genetic variants and potential drug targets more accurately, enabling greater precision in personalized medicine that will optimize treatments for patients [52].

10.7 CONCLUSION AND FUTURE SCOPE

The role of genomics in drug discovery and clinical medicine has proved transformative, contributing to the advance of gene-editing, precision medicine, bioinformatics, and artificial intelligence. The advent of genomics-based research has changed the way in which diseases are understood, leading to more targeted and far more effective treatments. For example, CRISPR as a gene editing technology has the potential to change treatment paradigms for many genetic disorders, and personalized medicine tailors treatments based on an individual's genetics to improve therapeutic response while minimizing adverse effects. This has resulted in the ability of researchers to identify novel drug targets and disease-associated biomarkers facilitated by bioinformatics that helps in the management and analysis of huge amounts of genomic data. AI also shortens this cycle by allowing us to analyze data better, understand drug relevance predictions, and determine the most optimal path in drug discovery.

In the future, genomics in concert with both AI and bioinformatics will further propel medical science forward at an accelerated pace as this witnesses a growing number of drugs being developed faster, and more tailored interventions. The shift toward preventative healthcare, which is based on genetic risk factors, and the trend leading towards personalized treatment plans for complex diseases such as cancer, cardiovascular disorder, or neurological conditions, describes a landscape of clinical medicine very different from today's public health. Nevertheless, these technologies bring along significant challenges for responsible and equitable usage due to issues like ethical concerns, data privacy, and accessibility.

KEYWORDS

- **artificial intelligence**
- **bioinformatics**
- **drug discovery**
- **gene editing**
- **genomics**
- **personalized medicine**
- **pharmacogenomics**

REFERENCES

1. Shen, R., Shen, X., & Wu, P. (2024). Economic impact of gene editing technology: The contribution of personalized medicine, bioinformatics, and artificial intelligence. *Journal of Simulation, 12*(1), 1.
2. Yildirim, Z., Swanson, K., Wu, X., Zou, J., & Wu, J. (2025). Next-gen therapeutics: Pioneering drug discovery with iPSCs, genomics, AI, and clinical trials in a dish. *Annual Review of Pharmacology and Toxicology, 65*. https://doi.org/10.1146/annurev-pharmtox-102522-023435 (accessed on 5th January 2026).
3. Singh, B., & Kaunert, C. (2024). Adventure in high altitude of mountainous topographies and health impacts: Lensing tourism sustainability via reducing ecological and sociocultural footprint and health emergency and medical assistance management. In G. Meraj, S. Hashimoto, & P. Kumar (Eds.), *Navigating Natural Hazards in Mountainous Topographies* (p. 15). *Disaster Risk Reduction*. Cham: Springer. https://doi.org/10.1007/978-3-031-65862-4_15 (accessed on 5th January 2026).
4. Sindelar, R. D. (2024). Genomics, other "OMIC" technologies, precision medicine, and additional biotechnology-related techniques. In *Pharmaceutical Biotechnology: Fundamentals and Applications* (pp. 209–254). Cham: Springer International Publishing.
5. Singh, B., Singh, A., Kaunert, C., Arora, M. K., Lal, S., & Ravesangar, K. (2025). Unleashing the ethical and legal implications of e-business revolution: Consumer privacy and security concerns in phase of digital disruption. In H. Taherdoost, G. Drazenovic, M. Madanchian, I. Khan, & O. Arshi (Eds.), *Business Transformation in The Era of Digital Disruption* (pp. 157–180). IGI Global. https://doi.org/10.4018/979-8-3693-7056-8.ch006 (accessed on 5th January 2026).
6. Udegbe, F. C., Ebulue, O. R., Ebulue, C. C., & Ekesiobi, C. S. (2024). Precision medicine and genomics: A comprehensive review of IT-enabled approaches. *International Medical Science Research Journal, 4*(4), 509–520.
7. Hashim, J. (2024). Genomic medicine from bench to bedside in clinical practice. *International Journal of Advanced Engineering Technologies and Innovations, 1*(4).
8. Alex, K., & Winkler, E. C. (2024). Comparative ethical evaluation of epigenome editing and genome editing in medicine: First steps and future directions. *Journal of Medical Ethics, 50*(6), 398–406.
9. Singh, B., & Kaunert, C. (2025). Innovative business models in digital and virtual reality for the gig economy: Space-age sustainable development projecting smart business practices. In H. Taherdoost, G. Drazenovic, M. Madanchian, I. Khan, & O. Arshi (Eds.), *Business Transformation in The Era of Digital Disruption* (pp. 127–156). IGI Global. https://doi.org/10.4018/979-8-3693-7056-8.ch005 (accessed on 5th January 2026).
10. Liu, F., Li, R., Zhu, Z., Yang, Y., & Lu, F. (2024). Current developments of gene therapy in human diseases. *MedComm, 5*(9), e645.
11. Dimitrievska, M., Bansal, D., Vitale, M., Strouboulis, J., Miccio, A., Nicolaides, K. H., El Hoss, S., Shangaris, P., & Jaców-Malinowska, J. (2024). Revolutionizing healing: Gene editing's breakthrough against sickle cell disease. *Blood Reviews*, 101185.

12. Kartha, R. V., & Orchard, P. J. (2024). Development of gene therapies from an academic perspective. In *Development of Gene Therapies* (pp. 351–368). Chapman and Hall/CRC.
13. Villagómez-Guzmán, A. K., & Quiroz, I. V. (2024). Human diseases and recent biotechnology breakthroughs in curbing diseases. *Biotechnology and Drug Development for Targeting Human Diseases, 9*, 165–187.
14. Singh, B., Malviya, R., & Kaunert, C. (2024). Elevating workplace sustainability for employees lensing mental health advancements: Runway for future ready healthcare services projecting SDG 3 (good health and well-being). In E. Shaikh (Ed.), *Impact of Corporate Social Responsibility on Employee Wellbeing* (pp. 285–310). IGI Global. https://doi.org/10.4018/979-8-3693-3470-6.ch014 (accessed on 5th January 2026).
15. Marques, L., Costa, B., Pereira, M., Silva, A., Santos, J., Saldanha, L., Silva, I., Magalhães, P., Schmidt, S., & Vale, N. (2024). Advancing precision medicine: A review of innovative *in silico* approaches for drug development, clinical pharmacology and personalized healthcare. *Pharmaceutics, 16*(3), 332. https://doi.org/10.3390/pharmaceutics16030332 (accessed on 5th January 2026).
16. Balogun, O. D., Ayo-Farai, O., Ogundairo, O., Maduka, C. P., Okongwu, C. C., Babarinde, A. O., & Sodamade, O. T. (2024). The role of pharmacists in personalised medicine: A review of integrating pharmacogenomics into clinical practice. *International Medical Science Research Journal, 4*(1), 19–36.
17. Udegbe, F. C., Ebulue, O. R., Ebulue, C. C., & Ekesiobi, C. S. (2024). Synthetic biology and its potential in US medical therapeutics: A comprehensive review: Exploring the cutting-edge intersections of biology and engineering in drug development and treatments. *Engineering Science & Technology Journal, 5*(4), 1395–1414.
18. Singh, B., Kaunert, C., & Malviya, R. (2024). Featuring machine learning in municipal solid waste management: Technological advancements in waste sorting and processing ropeway for SDG 13 (climate action). In R. Mehra, P. Kumar, & B. Hung (Eds.), *Municipal Solid Waste Management and Recycling Technologies* (pp. 343–366). IGI Global. https://doi.org/10.4018/979-8-3693-4054-7.ch013 (accessed on 5th January 2026).
19. Abrahams, E., & Downing, G. J. (2024). On the modern evolution of personalized medicine. In *The New Era of Precision Medicine* (pp. 1–25). Academic Press.
20. Singh, B., & Kaunert, C. (2024). Transforming smart city and smart sustainable tourism projecting artificial intelligence and IoT: Immersive potential for revolutionizing urban living and enhancing travel experience. In K. Jermsittiparsert & P. Suanpang (Eds.), *Technological and Managerial Approaches to Fostering Sustainable Travel* (pp. 55–80). IGI Global. https://doi.org/10.4018/979-8-3693-8115-1.ch003 (accessed on 5th January 2026).
21. Raghav, A., Singh, B., Jermsittiparsert, K., Raghav, R., & Yadav, U. (2024). Artificial intelligence in environmental and climate changes: A sustainable future. In B. Singh, C. Kaunert, K. Vig, & S. Dutta (Eds.), *Maintaining a Sustainable World in the Nexus of Environmental Science and AI* (pp. 485–506). IGI Global. https://doi.org/10.4018/979-8-3693-6336-2.ch019 (accessed on 5th January 2026).
22. Kaushik, T. K., Ravish, Singh, A., Raghav, A., Singh, B., Arora, M. K., & Lal, S. (2024). Untapping artificial intelligence in environmental and sustainable progression: Projecting legal and ethical considerations. In B. Singh, C. Kaunert, K. Vig, & S. Dutta

(Eds.), *Maintaining a Sustainable World in The Nexus of Environmental Science and AI* (pp. 109–130). IGI Global. https://doi.org/10.4018/979-8-3693-6336-2.ch005 (accessed on 5th January 2026).

23. Adegunsoye, A., Kropski, J. A., Behr, J., Blackwell, T. S., Corte, T. J., Cottin, V., Glanville, A. R., Glassberg, M. K., Griese, M., Hunninghake, G. M., & Johannson, K. A. (2024). Genetics and genomics of pulmonary fibrosis: Charting the molecular landscape and shaping precision medicine. *American Journal of Respiratory and Critical Care Medicine, 210*(4), 401–423.
24. Mackall, C. L., Bollard, C. M., Goodman, N., Carr, C., Gardner, R., Rouce, R., Sotillo, E., Stoner, R., Urnov, F. D., Wayne, A. S., & Park, J. (2024). Enhancing pediatric access to cell and gene therapies. *Nature Medicine*, *17*, 1–1.
25. Singh, B., Kaunert, C., Malviya, R., Lal, S., & Arora, M. K. (2024). Scrutinizing consumer sentiment on social media and data-driven decisions for business insights: Fusion of artificial intelligence (AI) and business intelligence (BI) foster sustainable growth. In A. Natarajan, M. Galety, C. Iwendi, D. Das, & A. Shankar (Eds.), *Intersection of AI and Business Intelligence in Data-Driven Decision-Making* (pp. 183–210). IGI Global. https://doi.org/10.4018/979-8-3693-5288-5.ch007 (accessed on 5th January 2026).
26. Zhang, Y., Yu, J., Xie, X., Jiang, F., & Wu, C. (2024). Application of genomic data in translational medicine during the big data era. *Frontiers in Bioscience-Landmark, 29*(1), 7.
27. Singh, B., Kaunert, C., Jermsittiparsert, K., Lal, S., Arora, M. K., & Raghav, A. (2024). Ultimate deep learning acclimatization with artificial intelligence drones for environment shield and climate change: Lifting sustainability towards fostering SDG 13. In B. Singh, C. Kaunert, K. Vig, & S. Dutta (Eds.), *Maintaining a Sustainable World in The Nexus of Environmental Science and AI* (pp. 59–80). IGI Global. https://doi.org/10.4018/979-8-3693-6336-2.ch003 (accessed on 5th January 2026).
28. Singh, B., Kaunert, C., & Gautam, R. (2024). Harnessing machine learning intelligence against cyber threats: Navigating legal and ethical challenges in global cyber investigations. In F. Özsungur (Ed.), *Empowering Entrepreneurial Mindsets with AI* (pp. 319–352). IGI Global. https://doi.org/10.4018/979-8-3693-7658-4.ch016 (accessed on 5th January 2026).
29. Jain, M. S., Srikruthi, K. S., Goudanavar, P., & Naveen, N. R. (2024). Navigating the frontier: Comprehensive insights into CRISPR technology advancements, delivery strategies, and ethical considerations in cancer research. *Oral Oncology Reports*, *22*, 100224.
30. Singh, B., & Kaunert, C. (2024). AI-driven strategies for customer engagement, market segmentation, and resource optimization: Projecting end user satisfaction and futuristic growth of business. In V. Nadda, P. Tyagi, A. Singh, & V. Singh (Eds.), *AI Innovations in Service and Tourism Marketing* (pp. 104–128). IGI Global. https://doi.org/10.4018/979-8-3693-7909-7.ch006 (accessed on 5th January 2026).
31. Aljabali, A. A., El-Tanani, M., & Tambuwala, M. M. (2024). Principles of CRISPR-Cas9 technology: Advancements in genome editing and emerging trends in drug delivery. *Journal of Drug Delivery Science and Technology*, *6*, 105338.
32. Singh, B., & Kaunert, C. (2025). Dynamic landscape of artificial general intelligence (AGI) for advancing renewable energy in urban environments: Synergies with

SDG 11—sustainable cities and communities lensing policy and governance. In S. El Hajjami, K. Kaushik, & I. U. Khan (Eds.), *Artificial General Intelligence (AGI) Security*. Springer. https://doi.org/10.1007/978-981-97-3222-7_12 (accessed on 5th January 2026).

33. Singh, B., & Kaunert, C. (2024). Blue ocean and machine learning trajectories SDG 14: Life below water for handling ocean pollution: Metaverse conserve ocean health sustainability through the lens of transboundary legal-policy regulations as articulating space for futuristic changes. In D. De, D. Sengupta, & T. A. Tran (Eds.), *Artificial Intelligence and Edge Computing for Sustainable Ocean Health*. Springer. https://doi.org/10.1007/978-3-031-64642-3_4 (accessed on 5th January 2026).
34. Dutta, P. K., Singh, B., Towfeek, A. S., Adamopoulou, J. P., Bardavouras, A. N., Bamwerinde, W., et al. (2024). IoT revolutionizes humidity measurement and management in smart cities to enhance health and wellness. *Mesopotamian Journal of Artificial Intelligence in Healthcare*, 110–117. https://doi.org/10.58496/MJAIH/2024/013 (accessed on 5th January 2026).
35. Singh, B., Kaunert, C., Malviya, R., Gautam, R., & Singh, G. (2024). Generative artificial intelligence in the tourism-hospitality industry: Strengthening customer relationships, delivering better experiences. In M. Talukder, S. Kumar, & P. Tyagi (Eds.), *Hotel and Travel Management in the AI Era* (pp. 275–300). IGI Global. https://doi.org/10.4018/979-8-3693-7898-4.ch013 (accessed on 5th January 2026).
36. Singh, B., & Kaunert, C. (2024). FedHealth in wearable healthcare, orchestrated federated deep learning for smart healthcare: Health monitoring and healthcare informatics lensing challenges and future directions. In *Federated Deep Learning for Healthcare* (pp. 207–223). CRC Press. https://www.taylorfrancis.com/chapters/edit/10.1201/9781032694870-16/fedhealth-wearable-healthcare-orchestrated-federated-deep-learning-smart-healthcare-bhupinder-singh-christian-kaunert (accessed on 5th January 2026).
37. Xu, F., Zheng, C., Xu, W., Zhang, S., Liu, S., Chen, X., & Yao, K. (2024). Breaking genetic shackles: The advance of base editing in genetic disorder treatment. *Frontiers in Pharmacology, 15*, 1364135.
38. Singh, B. (2024). Digital twins as new paradigm for bridging the gap from personalized medicine to specific rural public health: Exploring foundations, legal angels, and technologies for Health 5.0—Future research directions. In *Artificial Intelligence Solutions for Cyber-Physical Systems* (pp. 286–309). Auerbach Publications. https://www.taylorfrancis.com/chapters/edit/10.1201/9781032694375-16/digital-twins-new-paradigm-bridging-gap-personalized-medicine-specific-rural-public-health-bhupinder-singh (accessed on 5th January 2026).
39. Singh, B., & Kaunert, C. (2024). Hitching artificial intelligence and IoT and for livestock management competence: Line up nature and technology for smart agriculture. In M. Zia Ul Haq & I. Ali (Eds.), *Revolutionizing Pest Management for Sustainable Agriculture* (pp. 499–520). IGI Global. https://doi.org/10.4018/979-8-3693-3061-6.ch020 (accessed on 5th January 2026).
40. Singh, B., & Kaunert, C. (2024). Tagging blockchain technology fostering panacea for data privacy, cloud computing, and integrity: Legal framework in India and at a globe. In *Artificial Intelligence Solutions for Cyber-Physical Systems* (pp. 206–213). Auerbach Publications. https://www.taylorfrancis.com/chapters/

edit/10.1201/9781032694375-11/tagging-blockchain-technology-fostering-panacea-data-privacy-cloud-computing-integrity-bhupinder-singh-christian-kaunert (accessed on 5th January 2026).
41. Khan, I. R., Grover, V., Fatima, I., Alam, M., & Ahmad, N. (2024). The application of AI in drug discovery: Challenges, opportunities, and strategies. In *Analyzing Explainable AI in Healthcare and the Pharmaceutical Industry* (pp. 181–198).
42. Singh, B., & Kaunert, C. (2024). Redefining the nodes of metaverse and luxury hospitality: Unveiling AI and tech-driven solutions in hospitality and tourism industry. In M. Talukder, S. Kumar, & P. Tyagi (Eds.), *Impact of AI and Tech-Driven Solutions in Hospitality and Tourism* (pp. 175–202). IGI Global. https://doi.org/10.4018/979-8-3693-6755-1.ch010 (accessed on 5th January 2026).
43. Singh, B., & Kaunert, C. (2024). Trans-disciplinary collaborative disciplines generating cohesive curriculum: Runway for innovation, cooperation, and intentionality in higher education. In R. Kumar, E. Ong, S. Anggoro, T. Toh, & M. Fukui (Eds.), *Transdisciplinary Teaching and Technological Integration for Improved Learning: Case Studies and Practical Approaches* (pp. 152–176). IGI Global. https://doi.org/10.4018/979-8-3693-8217-2.ch008 (accessed on 5th January 2026).
44. Kumar, S., Malviya, R., Uniyal, P., Srivastava, S., Sridhar, S. B., Obaidur Rab, S., et al. (2024). Plant fiber composites for biomedical applications: Advances and prospective. *Current Pharmaceutical Biotechnology*, advance online publication. https://doi.org/10.2174/0113892010314833240815113622 (accessed on 5th January 2026).
45. Singh, B., & Kaunert, C. (2024). Marvel of biosensors in smart healthcare and application of internet of medical things for diagnosis: Unveiling benefits, challenges and futuristic approach. In *Smart Healthcare Systems* (pp. 187–198). CRC Press. https://www.taylorfrancis.com/chapters/edit/10.1201/9781032698519-14/marvel-biosensors-smart-healthcare-application-internet-medical-things-diagnosis-bhupinder-singh-christian-kaunert (accessed on 5th January 2026).
46. Singh, B., Kaunert, C., & Jermsittiparsert, K. (2025). Advocating medical tourism and healthcare in global health marketplaces: Promoting sustainability and technological advancements. In K. Jermsittiparsert & R. Marzo (Eds.), *AI Technologies and Advancements for Psychological Well-Being and Healthcare* (pp. 327–350). IGI Global. https://doi.org/10.4018/979-8-3693-9158-7.ch012 (accessed on 5th January 2026).
47. Singh, B., & Kaunert, C. (2025). Blockchain framework for precision medicine, clinical trials, and genomic biomarkers discovery: Focuses on monitoring and managing data in health dimensions. In K. Jermsittiparsert & R. Marzo (Eds.), *AI Technologies and Advancements for Psychological Well-Being and Healthcare* (pp. 255–288). IGI Global. https://doi.org/10.4018/979-8-3693-9158-7.ch010 (accessed on 5th January 2026).
48. Olono, A., Mitesser, V., Happi, A., & Happi, C. (2024). Building genomic capacity for precision health in Africa. *Nature Medicine, 30*(7), 1856–1864.
49. Singh, N., Lal, S., Raghav, A., Arora, M. K., Singh, B., & Kaunert, C. (2024). Power of ML and DL in precision modeling and climate predictions: Next-gen ocean insights. In B. Riswandi, B. Singh, C. Kaunert, & K. Vig (Eds.), *AI Applications for Clean*

Energy and Sustainability (pp. 254–273). IGI Global. https://doi.org/10.4018/979-8-3693-6567-0.ch013 (accessed on 5th January 2026).

50. Singh, B., Kaunert, C., & Kaushik, T. K. (2024). Unscrambling financial fraud with AI and machine learning in e-commerce transactions: Airing into ad clicks, credit card management. In B. Pandow, F. Masoodi, J. Iqbal, & G. Hussain (Eds.), *Navigating the Future of Finance in The Age of AI* (pp. 253–271). IGI Global. https://doi.org/10.4018/979-8-3693-4382-1.ch013 (accessed on 5th January 2026).
51. Yadav, S., Singh, A., Singhal, R., & Yadav, J. P. (2024). Revolutionizing drug discovery: The impact of artificial intelligence on advancements in pharmacology and the pharmaceutical industry. *Intelligent Pharmacy, 24.*
52. Huang, D., Yang, M., Wen, X., Xia, S., & Yuan, B. (2024). AI-driven drug discovery: Accelerating the development of novel therapeutics in biopharmaceuticals. *Journal of Knowledge Learning and Science Technology, 3*(3), 206–224.

Index

D

E

J

K

L

M

N

O

P

S

T

Y

Z